MW01674718

An Integrated Approach to Environmental Management

An Integrated Approach to Environmental Management

Evelyn Robinson

STATES
ACADEMIC PRESS
www.statesacademicpress.com

An Integrated Approach to Environmental Management
Evelyn Robinson
ISBN: 978-1-63989-044-6 (Hardback)

© 2021 States Academic Press

STATES
ACADEMIC PRESS

Published by States Academic Press,
109 South 5th Street,
Brooklyn, NY 11249, USA

Cataloging-in-Publication Data

An integrated approach to environmental management / Evelyn Robinson.
p. cm.
Includes bibliographical references and index.
ISBN 978-1-63989-044-6
1. Environmental management. 2. Environmental protection. 3. Environmental sciences. 4. Management.
I. Robinson, Evelyn.
GE300 .I58 2021
363.705--dc23

This book contains information obtained from authentic and highly regarded sources. All chapters are published with permission under the Creative Commons Attribution Share Alike License or equivalent. A wide variety of references are listed. Permissions and sources are indicated; for detailed attributions, please refer to the permissions page. Reasonable efforts have been made to publish reliable data and information, but the authors, editors and publisher cannot assume any responsibility for the validity of all materials or the consequences of their use.

Trademark Notice: All trademarks used herein are the property of their respective owners. The use of any trademark in this text does not vest in the author or publisher any trademark ownership rights in such trademarks, nor does the use of such trademarks imply any affiliation with or endorsement of this book by such owners.

For more information regarding States Academic Press and its products, please visit the publisher's website www.statesacademicpress.com

Table of Contents

Preface

Managing the impact of human societies on the environment is termed as environmental management. It focuses on ensuring that ecosystem services are protected and maintained for future human generations. It also aims to preserve the integrity of the ecosystem while considering the ethical, economic and scientific variables. Environmental management involves managing all the components of the biophysical environment, both living and non-living as well as the relationships among all species and their habitats. It attempts to identify factors which are affected by the conflicts arising between human needs and protection of natural resources. Environmental management can also be achieved by living sustainably, through systematic changes to transport, urban planning, individual lifestyles and consumerism. While understanding the long-term perspectives of the topics, the book makes an effort in highlighting their impact as a modern tool for the growth of the discipline. It studies, analyses and upholds the pillars of environmental management and its utmost significance in modern times. This book is a complete source of knowledge on the present status of this important field.

A detailed account of the significant topics covered in this book is provided below:

Chapter 1- The set of activities which seek to improve and maintain the state of an environmental resource which is impacted by human activities are known as environmental management. It involves environmental auditing and natural resource management. This chapter will provide a brief introduction to all the significant aspects of environmental management.

Chapter 2- The contamination of the biological or physical aspects of the earth to a level where basic environmental processes are affected is termed as environmental pollution. It can be broadly categorized as water pollution, air pollution and soil pollution. This chapter discusses in detail these types of environmental pollution.

Chapter 3- Environmental waste encompasses different categories of waste such as hazardous waste, radioactive waste, municipal solid waste, e-waste and biomedical waste. Its management involves the processes which are responsible for handling and disposing off this waste safely. The topics elaborated in this chapter will help in gaining a better perspective about these types of environmental waste and their management.

Chapter 4- The laws which seek to protect the environment are termed as environmental laws. A few types of environmental laws are air quality laws, water quality laws and waste management laws. This chapter has been carefully written to provide an easy understanding of the varied facets of environmental laws and policies such as environmental governance, environmental impact assessment and environmental globalization.

Chapter 5- The database and system which incorporates processes and procedures to train personnel, monitor, summarize and report specialized environmental performance information to the external as well as the internal stakeholders of an organization is known as an environmental management system. This chapter discusses the different aspects of environmental management systems in detail.

I would like to make a special mention of my publisher who considered me worthy of this opportunity and also supported me throughout the process. I would also like to thank the editing team at the back-end who extended their help whenever required.

Evelyn Robinson

1

Introduction to Environmental Management

The set of activities which seek to improve and maintain the state of an environmental resource which is impacted by human activities are known as environmental management. It involves environmental auditing and natural resource management. This chapter will provide a brief introduction to all the significant aspects of environmental management.

The ecological balance of our planet has been disturbed through a technological revolution, heavy industrialization, growth in transportation, unplanned urbanization and of course exploitation of resources. Ecosystem stability is in danger as the relationship between human beings and the environment has deteriorated. The relationship can once again become harmonious if people start encouraging activities like conservation, regeneration, and protection of nature.

Environmental management is concerned with marine, land and atmospheric conditions and works towards issues like deforestation and global warming. It looks at the carbon footprint and tries to find ways to reduce the irreversible damage people are leaving behind. Environmental management is described as a process to minimize waste and maximize compliance. It is created to address the environmental issues that have a direct and indirect impact on the globe and is affecting it adversely. It deals in finding appropriate solutions to environmental crises and preventing ecological disasters. Environmental management also investigates potential sources of renewable energy so that fossil fuel does not become depleted.

Objectives of Environmental Management

The objectives of environmental management are as follows:

- Identifying environmental issues.
- Finding solutions for environmental issues.
- Establishing limits to avoid overuse.
- Help to renew natural resources.
- Minimize the use of natural resources.
- Developing monitoring systems and research institutions.
- Regeneration of degraded environment.
- Review the environmental goals of an organization.
- Setting environmental targets to minimize the environmental impact of an organization.
- Control environmental pollution.

- Ensuring environmental awareness program is being followed by every employee.
- Review existing technologies and try to make then eco-friendly.
- Make maximum utilization of natural resources.
- Assess the impacts of potential activities on the environment.
- Encourage resource conservation programs.
- Develop strategies for improving the quality of life.
- Implement ways for environmental protection.
- Minimize the impact of natural disasters.
- Identify, develop and implement policies related to sustainable development.

Importance of Environmental Management

Times are changing and so is the mindset of people. There was a time when business entities were only concerned about sales figures and revenues. In recent times there has been an outcry about the impact of business activities on nature and how the resources are being depleted and compromised.

People have become aware of the environment and are looking to deal with those companies or groups that are associated with environmental management concepts. In order to move with the new trends and mindset, most firms are implementing effective environmental management programs that work towards conserving and protecting resources as well as the environment.

Environmental management has an important role to play in the protection of our environment. Customers, clients, employees, stakeholders, investors expect business entities to manage their activities in such a manner that it does not have a negative impact on the environment. It has become necessary for every business entity to implement an environmental management system in place to manage environmental risks. The importance of having an effective environmental management process in place is as follows:

- Reducing carbon emissions.
- Processing all the waste in a safe and effective manner.
- The effective and wise use of energy and resources.
- Preventing pollution.

Features of Environmental Management

- The concept of environmental management acts as a tool to improve environmental performance.
- It works towards waste minimization.
- Is a systematic approach for overseeing environmental affairs of an organization.

- Addresses concerns related to the environment through the evaluation of related practices.
- Encourages employees to undergo training to achieve desired objectives.
- Helps to determine the impact and objective of a product or service.
- Assigns responsibility and accountability in a firm.

Advantages of Environmental Management

- Effective use of resources: Organizations have developed and executed several policies to manage resources, minimize waste and reduce costs.
- Improve sustainability: Effective environmental management policies encourage business entities to improve sustainability so that they can gain a competitive edge and give the business a good chance for long-term success.
- Helps in staying out of trouble: An organization that implements environmental management policies stays ahead of regulations imposed by environmental affairs division. He does not have to pay any penalty for violating the balance of nature or adopting unethical practices.
- Pollution control: Environmental management practice has encouraged numerous business entities to take steps for pollution control.
- Improved public image: An organization with environmental management procedures in place works towards preventing pollution, minimizing environmental risks and preserving natural resources. This will enhance the trust of its customers, attract new clients, boost employee morale and help the firm in entering new markets.
- Economic progress: The environmental management theory encourages business entities to participate in social welfare programs to enable economic progress of a nation.
- Finance: Because of their environmental management policies and procedures companies are able to impress investors. It has now become easier to get the necessary finance.
- Improved management efficiency: An integrated environmental management policy will improve the overall management system of a firm. Documenting routine operations and compliance plans will help in capturing the attention of other employees and streamlining operations. This will help to decrease waste and environmental risk.
- Better compliance: Implementing an environmental management process ensures that an organization is fulfilling its responsibility towards a safe environment.
- Marketing: Business entities are projecting themselves as companies that are eco-friendly. They are highlighting their commitment to improved environmental policies via impactful advertising and this has helped them in creating a strong customer base.
- Sales opportunities: People are becoming eco-friendly and want to be associated with like-minded companies. The organization which promotes its products and services as eco-friendly has increased sales opportunities.

- Empowered employees: Employees can relate to environmental management program and understand its impact on the society as a whole. This encourages employee empowerment as they believe it is their contribution that is making a difference in the world.
- Certification of recognized standards: When a company strives to obtain a certification to recognized standards it is gaining business credibility with stakeholders and customers alike.

Disadvantages of Environmental Management

- Cost: The cost of developing, monitoring and implementing effective environmental management policies is high. It becomes difficult for every business entity to implement such programs in its business plan because of the high cost.
- Too much of trouble: The environmental management system can prove too much of a trouble if you are a start-up or small business entity who have limited resources at its disposal.
- Time and resources: The objective of implementing environmental management policies is to minimize energy and unnecessary resource consumption and reduce waste. It is a bitter fact that every such scheme requires additional use of time and resources that can prove expensive in terms of time, effort and resources to an organization.
- Training costs: In this process that will need recruiting human resources for all its stages. An organization has to offer training to its recruits if it wants to implement an effective system and this can prove expensive and troublesome to management.

Public Awareness in Environmental Management

Quality and integrity of environment can be preserved by public awareness. Dwindling environment is the biggest threat to the existence of human beings. Loss of valuable natural resources and pollution lead to harmful effects as well as abiotic stress on flora and fauna. Public awareness is one of the basic principles in environment management. It involves:

- Developing sensitivity and awareness towards environmental issues.
- Inculcating and imbibing the stringent need for conservation of natural resources and surrounding environment.
- Encouraging active participation of the people in environmental protection and development.
- Developing expertise for the active identification and finding remedy to environmental problems.
- Developing expertise for the active identification and finding remedy to environmental problems.
- Assessment of socio-economic, ecological and aesthetic aspects and impacts of various environmental programmes initiated to improve environmental health.

Need of Public Awareness in Environment Management

It is very important to make public aware about the deadly consequences of the environmental degradation as it would lead to massive extinction of life if left unattended and reformative measures are not undertaken. We are facing various environmental challenges that need to be tackled stringently for sustainable growth and development. Eco-friendly approach needs to be followed in every nation to acquaint for the threats posed in the name of industrialization and development. Natural resources are limited in the world. We are dependent on natural ecosystems for the products obtained from forests, grasslands, oceans and from agriculture and livestock as well as water, air, soil, minerals, oil etc. which are indispensable part of our life support systems. Life would be impossible without these all substances. Increase in population put pressure on these limited natural resources.

The earth cannot sustain the ever growing demand for resources. Moreover misuse of resources is the other contributing factor to environmental deterioration. Wastage and pollution of water resources, generation of non-biodegradable materials like plastic, non-recyclable electronic waste, nuclear waste are the other serious threats. Manufacturing processes generate solid waste, chemicals and gases that pollute the environment. Alarming increase in waste generation cannot be managed by natural processes as most of the synthetic waste is non-biodegradable. These keep on accumulating in our environment leading to a variety of diseases and other adverse environmental effects that seriously affects our lives. Air pollution leads to chronic respiratory diseases, water pollution to gastro-intestinal diseases, and many toxic pollutants are known to cause cancer. This alarming situation can only be dealt by initiatives taken by each one of the individuals daily towards preserving environmental resources. Neither government alone can manage and safeguard the environment, nor bunch of environmentalists can prevent environmental degradation. Sensitization can only be done to create self-responsibility. Therefore, efforts at individual level can lead to sustainable development.

Various Environmental Challenges

Population Explosion

Current population has reached nearly 1.34 billion with population growth rate of 1.2%. Some of the reasons for rapidly growing population are illiteracy, poverty, high fertility rate, reduced mortality rates and infiltration from neighboring countries. It puts extensive pressure on natural resources to meet growing demands. Hence, population growth is the biggest challenge. Women literacy and awareness are some of the measures that need to rigorously practice to circumvent this challenge.

Poverty

Environment poverty nexus dates back from ages. Poverty was one of the main agenda for the achievement of Millennium Development Goals (MDGs) framed by United Nations and its eradication from the world is the first goal for year 2030 Agenda for sustainable development. Although number of poor as well as the proportion of poor below the poverty line (according to the Tendulkar poverty line) have been declining over two decades as per government reports, but about 270 million are still below the poverty line. Initiatives like MGNREGA have been taken to improve the poverty presiding. Majority of poor people are directly dependent on the natural resources for their basic needs like food, fodder fuel and shelter. Environment degradation has adversely affected the status of poor who depend upon natural resources for their immediate needs. Thus, the challenge of poverty and the challenge of environment degradation are two sides of the same coin. The population growth and poverty are interlinked as every child is the bread winner for the family.

Agricultural Growth

The main aim of green revolution was making nation food sufficient but the strategies adopted in achieving this aim lead to harmful effects directly on the rural and indirectly on the urban communities. Introduction of high yielding varieties, indiscriminate use of fertilizers, insecticides have led to development of resistant varieties of flora and fauna, decreased soil fertility, disturbance of the natural ecosystems etc. Lack of appropriate knowledge to use insecticides and fertilizers, awareness for the banned chemicals and ignorance of the potential ill effects has led to surfacing of serious diseases like cancer along with harmful effects and disruption of agricultural ecosystems. The people must therefore, be acquainted with the eco-friendly methods to sustain and increase agricultural growth without damaging the environment.

Ground Water

Rational use of groundwater is necessary as aquifers are limited, confined and ground water level is declining day by day. Leaching chemical fertilizers and pesticides, industrial effluents have polluted surface water and affected the quality of the groundwater. It is essential to maintain the water quality of rivers and other water bodies. Strategies for provision of safe drinking water and keeping water bodies clean are the key challenges. Initiatives under Corporate social responsibility (CSR) by the industries for effluent treatment can be helpful. Mandatory and stringent law for tackling industries not abiding the rules for waste disposal and awareness among public to act as whistle blower can help to open new paths for providing fresh and pure drinking water to the nation.

Deforestation

Massive deforestation is the result of industrialization. Submerging of forests due to building of dams for hydroelectric power plants has led to displacement of local people, and damage to flora and fauna. Dams on river Narmada, Bhagirathi etc. have become hot topics for debate in political and scientific spheres. Forests cover has been decreasing from last few decades owing to pressures of development. Vast areas have turned into wastelands. These areas can be rejuvenated by increasing vegetative cover. The tribal communities respect flora and fauna to sustain their lives. Local people should be integrated with knowledge and skills of the forest department for restoring and conserving forests. Well planned strategies by joint management of forests should be evolved. We must recognize the power of awareness and integrity of the public towards environment issues from past. 'Narmada bachaoandolan' is one such acclaimed protest for preserving environmental integrity by the local people. The demonstrations, hunger strikes, protests, rallies and written representations have made an impact on the direction of the movement to stop the building of large and small dams along the Narmada. Media attention of these events has taken the issues from a local level to a more national scale. These protests have forced the World Bank to withdraw its loan from the projects.

Land Degradation

Land is a limited resource. With the increase in human population and economic development, demand for land is increasing. Total land area of India is 329 million hectares out of which only 266 million hectares have potential productivity. 143 million hectares of land is under cultivation and remaining land area (85 million hectares) has suffered from soil degradation. Nearly 30% of the land is degraded. From 123 million hectares, 40 million hectare land is completely unproductive. Overgrazing, water and soil erosion leads to further land degradation. This degradation can be avoided by reforestation and local public efforts to restore land usage.

Technological Advancement

Present day people are more concerned about using latest technologies without knowing potential ill effects of the gadgets and electronic devices. Although these have become indispensable part of living in present time but overuse and unnecessary ignorance about potential ill effects need to be tackled. Development has to go hand in hand with vigilance, sensitivity and security toward management of natural resources. This change can be brought in education, administrative procedures, outlook and institutions.

Genetic Diversity

Genetic diversity need to be conserved by taking proper measures. Interbreeding of the wild populations has greatly reduced with urbanization and deforestation leading to alarming disappearance wild genetic stocks/extinction. Lack of genetic diversity in tigers has been stated as the cause for their extinction. Habitat loss and fragmentation has led to confined breeding. The protected areas like sanctuaries, national parks, biosphere reserves are isolates populations thereby decreasing interbreeding efficiency. Steps need to be taken for preserving genetic biodiversity as every creature is important to maintain balance in nature.

Urbanization

Urbanization and industrialization has led to various environmental problems like solid waste disposal, development of slums and sewerage disposal. Clean towns and cities are therefore major challenge which can be accomplished only by the cumulative effort of government and people.

Air and Water Pollution

Most of the industrial setups do not comply with the environmental standards setup for waste disposal and treatment. Stringent implementation of the environmental policies is necessary to prevent environmental deterioration. Their implementation needs resources, technical expertise, political and social obligation. Awareness to environmental policies is mandatory for achieving public support for their implementation.

Environmental Auditing

Environmental auditing is a management tool which simply inspects the environmental management activities performed by the industries or organizations and makes them aware of new cleaner technology. For the impact of industries and their product on natural resources and environmental quality it is necessary to have "Environmental Audit" to ensure sustainable industrial developments.

Pollution now is an inevitable consequence of modern industrial technology, rapid and convenient transport and comfortable housing, but excessive pollution may interfere with man's health and his mental, social and economic wellbeing. There is now a clarion call from every nook and corner of the society that "save the nation from this menace the pollution". Governments, all over the world, have formulated laws and regulations to correct and cure the past violations of good environmental practices.

The term auditing is known to us in financial accounts and records are examined. Environmental audit is for the impact of the industries and their products on natural resources and environmental quality. It is necessary to have 'Environmental Audit' to ensure sustainable industrial developments. Environmental Audit is a pragmatic management tool, which addresses itself to help an industry or operation, to verify compliance with environmental requirements, to evaluate the effectiveness of the environmental management system, to assess risks and to identify and correct environmental hazards. It is the examination of accounts of revenues and costs of environmental and natural resources, their estimation, depreciations and natural resources, their estimation depreciations and values recorded in the books of accounts. Environmental organization management systems and equipment are performing with the aims of:

- Facilitating management control of environmental practices.
- Assessing compliance with company policies.
- Facilitating professional competence.

The concept of environmental auditing came into being in industrialized countries such as Canada, USA, UK and Netherland during early 1970's under a different number of approaches and names depending on the company concerned such as "Environmental review". Environmental performance appraisal, Quality approach, etc. In 1972, British Petroleum (B.P.) International group first used this concept. The United States Environmental Protection Agency published their environmental policy in 1986 followed by International Chamber of Commerce book let on environmental auditing in the year 1988.

Objectives of Environmental Auditing

- An environmental audit programme which is designed and implemented properly can enhance an industry's environmental performance.
- Monitoring the scale of optimum utilization of the resources and evaluating the company at national and international level.
- To suggest for using alternative energy for the conservation of energy resources.
- Evaluation of waste water quality and determination of waste water characteristics and their effects on the living system.
- Classification of the categories of solid waste hazardous waste their sources, quantities and characteristics.

- Introduction and implementation of time saving technologies in production.
- Maintains of Labour/Occupational health and medicine.
- Proper documentation of environmental compliance status.
- To help in minimizing the wastes through modern cleaner technologies.
- Regular environmental auditing once in a year will help in producing environmentally educated &technically sound personnels.

The purpose of the environmental audit is to provide an indication to the management of the improvements while environmental organization system and equipment are performing. To fulfil this purpose it is essential that audits should be seen as the responsibility of the company. The audit work can be voluntary and for the advantage of the company. The audit work can be done systematically and efficiently by the help of environmental auditing programme. It helps in the proper utilization of natural resources as a whole it improves environmental quality.

Environmental Audit Scheme and its Components

This particular tool is very important aspect of the environmental audit for the total management system in terms of its being an asset or a liability for the industry`s environmental performance. Environmental system is with a broad aim for a green environment.

- It helps in reducing waste.
- It helps in assessing compliance with regulatory requirement.
- It also helps in prevention control of effect of pollutant.
- It promotes relationship between qualified technician professionals, individual industries, State Pollution Control Board, other public authorities and industrial association etc.

Environmental Audit Scheme (EAS) has three following components:

- State Pollution Control Board: It plays active role in implementing the environmental audit effectively. The steps involved in state pollution control board are mentioned:
    - To prepare format of audit report on all the aspect of environmental protection.
    - To appoint internal auditors to prepare industries audit report.
    - Evaluation and verification of audit reports.
    - Initiating the action on evaluated report.
- Internal Auditor: The selection of auditor consist of experienced experts from various backgrounds. A qualified auditor should be requires as per the rules of State Pollution Control Board with well-equipped laboratory facility for analysis of water and air samples.
- External Auditor: Team should be approved by State Pollution Control Board. Evaluated and verified reports have to send their comments to State Pollution Control Board for further action.

Principle Areas of Environmental Auditing

- Material Audit: It mainly concentrates on the use of different raw material or natural use of resources, cost\unit, process wise consumption, wastage etc. Conservation of raw material, scientific storage and reuse of wastage material are taken into consideration.
- Energy Audit: It examines consumption of various forms of energy in different processes in any industry or organization. The main of audit is minimization, elimination of avoidable losses of valuable energy and their conservation.
- Water Audit: Consumption of water at different sources is noted. It also concentrates on the reuse and recycling of water, evaluation of raw water intake, balancing of water table and other sources.
- Health and Safety Audit: Workers and employee are the basic need of industry. Health and safety of those is well considered in audit. Proper disposal of toxic and hazardous waste, fire prevention measures etc. should be evaluated.
- Environmental Quality Audit: Conservation of every aspect and stage of environment helps to maintain the quality. This also well noted in audit scheme.
- Waste Audit: It covers the qualitative and quantities evaluation of waste generated from industries.
- Engineering Audit: Use of advance technology which will cover the suitable processes and engineering application.
- Compliance Audit: The different aspect of audits that are required to be carried out as per regulation, procedure and according to the policies of that particular industry are known as compliance audit.

Environmental Audit Activities

- Pre-audit activities: These include selection of the expert team and development of a plan. The plan should include defining the scope of the work, selecting priority areas, laying down the procedure and allocating team resources. There are four key activities:
    - Submitting pre-visit questionnaire of the facility.
    - Reviewing relevant regulation.
    - Defining audit scope and team responsibilities.
    - Reviewing audit check lists.
- On-site activities: Meeting of the team with the appropriate with the appropriate personnel of the unit and with discussion of the objective. The three primary functions on site activities are:
    - Record and documentation review.

    - Interview with staff.
    - Physical inspection of the facilities.
- Post-audit Activities: Development of raw material balance analysis for each process unit of the industry highlighting analysis for each process unit of the industry highlighting the proposed utilization of raw material and ideas regarding reuse can be thought of. Water, energy consumption, vegetation removal all the sensitive parts must be balanced. Evaluation of pollution audit, emission standard are carried out.
    - Issue of draft report.
    - Issue of final report.
    - Conducting follow up to the corrective action.

Advantages

- Preparation of Environmental management plan.
- Assessment of environmental input and risks.
- Identifying areas of strength and weakness for improvements.
- Evaluation of pollution control.
- Verification of compliance with laws.
- Assuring safety of plant, environment and human beings.
- Enhancement of loss prevention, manpower development and marketing.
- Budgeting for pollution control, waste prevention, reduction, recycling and reuse.
- Providing an opportunity for management to give credit for good environmental performance.
- As a whole environmental audit plays an important role in minimizing the environmental problem locally, regionally, nationally and internationally.

Natural Resource Management

Natural resource management refers to the management of resources such as animals, plants, soil, and land to improve the quality of life. It focuses on how they could possibly be managed for our today and tomorrow. It deals with the way in which people interact with natural landscapes. Specifically, it aims at understanding the scientific and technical aspects of resources and ecology. Environmental management is closely related to natural resource management.

An example of resource management is the lumber tree forest. These forests are less diverse because of the harvesting of lumber trees for a longer time. Such a forest is the result of the management of the resource by timber companies and can sustain life because the locations for harvest are rotated and the rate of tree cutting is controlled. The forest becomes a sustainable source of lumber for a long time period due to such practices.

Need to Manage Natural Resources

There is a need to manage our resources today, not just for us, but our future generations too. Natural resources are limited, there is a crisis for these resources, hence needs to be managed. For example, water: Usage of water and availability of water always needs to be balanced, with a growing population there is an exponential increase in the consumption of water while the availability of fresh water is decreasing by the day.

Natural resources are depleting every day not only because it is being used up by people, but also because it is being polluted and destroyed due to various human activities such as deforestation, modernization, global warming to name a few. There is an absolute need to manage resources for the existence of living beings on earth and their survival.

Objectives of Natural Resource Management

The objectives of natural resource management are as follows:

- To maintain ecological diversity.
- To provide resources for future generations.
- To maintain employment facilities for people.

Approaches to Management of Natural Resources

Now that we know how important natural resources are, what steps are we taking in the direction of saving ourselves from the crisis of natural resources? Are there any measures being taken at all? Let's find out in the upcoming segment. Natural resources need to be used in a sustainable manner for the welfare of mankind. They are so complex in nature that coming up with fresh frames as solutions seem to be the only way out.

- Various rules and regulations have been passed by the Government over the years to ensure that our wildlife does not become extinct and are not hunted down. The government has also passed rules and laws to limit large scale industries from the use of excess water and rules for proper sewage to ensure water bodies around industrial areas are not getting contaminated. Care has also been taken by the government with regard to air pollutants being released by industries in the air causing harmful diseases.
- Water is being treated regularly to provide drinkable water to the people.
- The government has adopted measures and encouraged people to recycle and reuse products as much as possible.
- Use of plastic has also been banned which reduces damage caused to nature to quite a large extent.

- Clean India Mission or Swachh Bharat Abhiyan was launched by PM Narendra Modi. It was initiated with the aim of having cleaner streets, society, and environment, contributing to the national and hence global welfare.

Management of natural resources focuses on preventing over-exploitation of resources. For e.g., with the advancement of agriculture practices, and farming have shifted from manual labour to the use of machines. The increasing use of pesticides has become damaging to the environment. All this has increased the need for natural resource management wisely.

References

- Environmental-management: marketing91.com, Retrieved 11, July 2020
- Environmental-Auditing-Its-Benefits-and-Counterance-235950655: researchgate.net, Retrieved 24, April 2020
- An-overview-of-natural-resource-management, biology: byjus.com, Retrieved 02, January 2020

2

Classification of Environmental Pollution

The contamination of the biological or physical aspects of the earth to a level where basic environmental processes are affected is termed as environmental pollution. It can be broadly categorized as water pollution, air pollution and soil pollution. This chapter discusses in detail these types of environmental pollution.

Water Pollution

Water is abundantly available in nature. It is an essential constituent of all animal and vegetable matter and form about 75% of the matter of Earth's crust. It is also an essential ingredient of animal and plant life. Water is distributed in nature in different forms, such as rain water, river water, spring water and underground in aquifers. Rain water is the purest form of naturally occurring water. Water pollution is the contamination of natural water present in lakes, rivers, streams, oceans, and groundwater due to inflow or deposition of pollutants directly or indirectly into the water systems. Any modifications or change in the chemical, physical and biological properties of water that can cause any harmful consequences on living things and the environment is known as water pollution. Water pollution is very often caused by human activities. In most of the developing countries the major cause of death is consumption of polluted water. Water pollution affects all living species, populations and the complete functioning of the ecosystem that lives in the waters.

Sources of Water Pollution

Classifications of water pollution according to identifiability, can be either a Point or a Nonpoint source. Point source refers to the pollutants that belong to a single identifiable source. The sources are called point sources because in mathematical modeling, they can be approximated as a mathematical point to simplify analysis. Nonpoint source pollution can include:

- On-site septic systems.
- Leaky tanks or pipelines containing petroleum products.
- Municipal landfills.
- Livestock wastes.
- Industrial/factory wastewater.
- Municipal sewage treatment plants.

Non-point source on the other hand are pollutants emitted from multiple scattered sources. Non-point sources of pollution are often termed 'diffuse' pollution and refer to those inputs and impacts which occur over a wide area and are not easily attributed to a single source. Nonpoint source pollution can include:

- Excess fertilizers, herbicides and insecticides from agricultural lands and residential areas.
- Oil, grease and toxic chemicals from urban runoff and energy production.
- Sediment from improperly managed construction sites, crop and forest lands, and eroding stream banks.
- Salt from irrigation practices and acid drainage from abandoned mines.
- Bacteria and nutrients from livestock, pet wastes and faulty septic systems.
- Atmospheric deposition and hydro modification.

Figure below depicts the different point and nonpoint sources leading to surface and groundwater pollution.

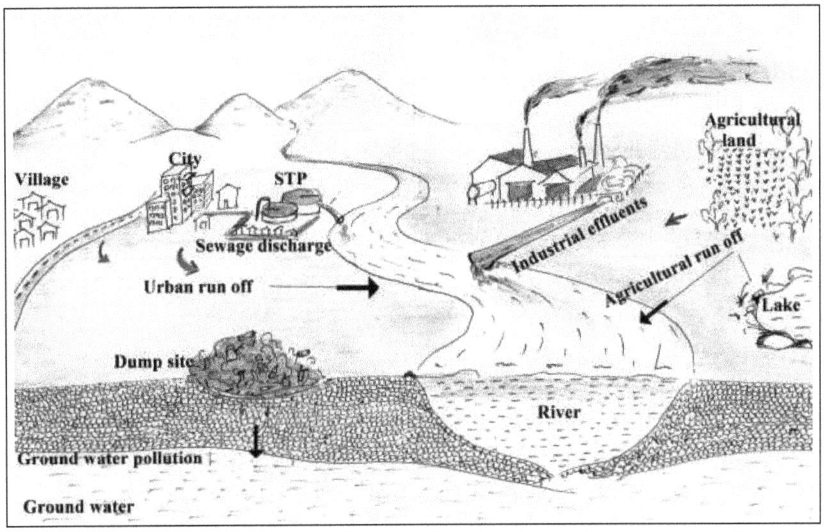

Examples of potential sources of contamination for surface and ground water.

Industries

Most industries generally produce wastes containing toxic heavy metals along with hazardous organic and inorganic effluents. These chemicals contaminate the ground water and severely pollute it. These pollutants ultimately find their way in effluents discharged by industries, which cause water pollution and damage to us and our environment. They contain pollutants such as lead, mercury, sulphur, asbestos, nitrates and many other harmful chemicals. Many industries do not have proper waste management system and drain the waste in the fresh water which goes into rivers, canals and later in to sea. The toxic chemicals have the capability to change the color of water, increase the amount of minerals, change the temperature of water and pose serious hazard to water organisms.

Sewage and Waste Water

Every day wastewater from sinks, dishwashers, clothes washers and toilets flows into sewer pipes containing pathogenic organisms, nutrients and solids from domestic wastes. Leakage from the sewer lines can contaminate the underground water and make it unfit for drinking. Also, when not repaired on time, the leaking water can come on to the surface and become a breeding ground for insects and mosquitoes. The sewage water carries harmful bacteria and chemicals that can cause serious health problems. Pathogens are known as a common water pollutant; the sewers of cities house several pathogens and thereby diseases. Microorganisms in water are known to be causes of some very deadly diseases and become the breeding grounds for other creatures that act like carriers. These carriers inflict these diseases via various forms of contact onto an individual. A very common example of this process would be malaria.

Table: Water borne pathogens and diseases caused by them.

Disease	Pathogen
Amoebiasis	Entamoeba histolytica
Giardiasis	Giardia lamblia
Botulism	Clostridium botulinum
Cholera	Vibrio cholerae
Dysentry	Shigella dysenteriae
Typhoid fever	Salmonella typhi
E. coli infection	Escherichia coli
Hepatitis A	Hepatovirus A
Polio	Poliovirus

Mining Activities

Mining is the process of crushing the rock and extracting coal and other minerals from underground. These elements when extracted in the raw form contain harmful chemicals and can increase the amount of toxic elements when mixed up with water, may result in health problems. Mining activities emit several metal waste and sulphides from the rocks and is harmful for the water. Acid mine drainage causes the acidity of water, which in turn disturb the chemical composition of water source by dissolving and freeing some harmful substances.

Agriculture Run-off

The water leaving from agricultural land is known as agricultural runoff. Runoff could be due to drainage by rain or by irrigation. Fertilizers, pesticides, insecticides, processing wastes and animal wastes etc. are constantly added to the water. Leachates from agricultural land containing nitrates, phosphates and potash, move horizontally as into surface water posing danger. For generations, thousands of acres of farmland have been cultivated throughout. Nutrient-rich water from farm fields is drained or pumped into natural water bodies without treatment which the impact water quality. This overabundance of nutrients encourages algal blooms i.e. overgrowth of aquatic weeds, that deplete oxygen from the water and block sunlight from reaching underwater vegetation, critical to fish and wildlife habitats. Nitrogen pollution comes from many sources, including from the fertilizers used on lawns and in landscaping. This nitrate-rich water makes its way to surface waters as runoff during rainfall or over-irrigation, or it may drain slowly from the soil over time.

Marine Dumping

In some countries the garbage produce by each household in the form of paper, aluminum, rubber, glass, plastic, food if collected and deposited into the sea. These items take from two weeks to 200 years to decompose. When such items enter the sea, they not only cause water pollution but also harm animals in the sea. Also waste from recreational activities and waste from travelling vessels are thrown into the oceans, causing ocean pollution.

Accidental Leakage and Marine Pollution

Oil spill pose a huge concern as large amount of oil enters into the sea which does not dissolve with water; there by opens problem for local marine wildlife such as fish, birds and sea otters etc. For example, a ship carrying large quantity of oil may spill oil if met with an accident and can cause varying damage to species in the ocean depending on the quantity of oil spill, size of ocean, toxicity of pollutant.

Burning of Fossil Fuels

Fossil fuels like coal and oil when burnt produce substantial amount of ash in the atmosphere. The particles which contain toxic chemicals (often of sulfur, which is present in coal) when mixed with water vapor result in acid rain. Also, carbon dioxide is released from burning of fossil fuels which result in global warming which in turn affect the global water cycle.

Facilities Handling Radioactive Materials

The element that is used in production of nuclear energy is Uranium which is a highly toxic chemical. Nuclear waste can have serious environmental hazards if not disposed-off properly. Few major accidents have taken place in Russia, USA and Japan. Nuclear research centers and medical sector also handles radioactive materials. From such sources negligence or accidental release of radioactive material causes water pollution. Mining also expose the environment to the radioactive materials.

Urban Development and Run-off

Growing population demands more for housing, food and cloth. As more cities and towns are developed, they have resulted in increased use of fertilizers to produce more food, soil erosion due to deforestation, increase in construction activities, inadequate sewer collection and treatment, landfills as more garbage is produced, increase in chemicals from industries to produce more materials. Runoff from urban areas contains large concentration of oils, greases, nutrients, heavy metals and detergents. The detergents being soluble can pass through the soil and pollute ground water. Raw sewage dumped in shallow soakpits and seepage from polluted lake, pond or stream also pollutes water.

Classification of Water Pollutants

Water pollutants can be divided into four major categories: Chemical, Physical, Physiological and Biological. Though the physical and physiological pollutions are mainly outcomes of chemical pollutants, they are categorized separately because of their nature.

Chemical Pollutants

Chemical pollutants can further be divided into Organic pollutants and Inorganic pollutants.

Organic Pollutants

Organic pollutants are the compounds those contain carbon atom in them and mainly of biological origin. These are the oxygen depleting substances in water. They are carried to the water course from different sources like sewage, animal and human excreta, refineries, distillery, pulp and paper industry, tanneries, dairy industry, textile industry, slaughter houses etc. BOD and COD is common indicator of high organic content of waste water. The organic pollutants could be further subdivided into two:

- Biodegradable organic pollutants: These are the organic materials which can be easily degraded by microorganisms. Most of the naturally occurring organic compounds (originated from plants and animals) are biodegradable. Examples are given in the table below.
- Non-biodegradable organic pollutants: These are the organic materials which are resistant to biological degradation. Most of these are synthetic organics, which are of high concern. Examples are given below in table. Some naturally occurring organic compounds biodegrade so slowly that they are considered refractory. For example, lignic acid, tannic acid, cellulose, and many of the organics associated with petroleum.

Table: Examples of some organic pollutants.

Types of Organic pollutants	
Biodegradable organic pollutants	Non-biodegradable organic pollutants
Proteins, fats, carbohydrates, alcohols, acids, aldehydes, esters, soaps etc.	Pesticides (DDT, aldrin, dieldrin, endrin, endosulfan, lindane, toxaphene, mirex, heptachlor etc.), polychlorinated biphenyls (PCB's), polyaromatic hydrocarbons (PAH's), plastics, dioxins, furans, phenols, dyes, detergents etc.

Inorganic Pollutants

Inorganic pollutants include inorganic acids, alkalis, salts, anions, cations, free chlorine, ammonia etc. They are added a result of industrial effluents, sewage, household cleansers and surface run-off from urban and agricultural field etc. They affect the physical and chemical quality of water. Some common inorganic pollutants are given below in table.

Table: Common inorganic pollutants found in water.

Pollutants	Examples
Acids	Phosphoric acid, sulfuric acid, hydrochloric acids etc.
Alkalis	Sodium hydroxide, lime etc.
Cations	Calcium, magnesium, sodium, potassium, ammonium, iron, manganese, aluminium, mercury, lead etc.
Anions	Phosphates, sulphates, chlorides, nitrites, nitrates, cyanides, carbonates, bicarbonates etc.

Physical Pollutants

These are the pollutants which deteriorate the physical parameters of water quality. These include color, turbidity, suspended matter, radioactivity, froth and thermal pollutants.

- Color: Color affects the aesthetic value of water body. It may or may not be harmful for health e.g. green color of water body due to humic acid (vegetable origin). Harmful color producing chemical includes various dyes, toxic colored compounds of iron and chromium etc. Color is also produced by pigments like that of microalgae.
- Turbidity: Turbidity is due to colloidal particles which do not settle in standing water. It may be due to presence of inert and harmless material like clay or due to presence of chemicals like hydrated iron oxide, aluminum hydroxide and organic substances etc. Greater turbidity means presence of greater amount of pollutants in water, but absence of turbidity does not mean that water is unpolluted.
- Suspended matter: These are coarse and insoluble matter suspended in water. Natural suspended matters which are not harmful for health include silt, clay, sand and rock particles etc. The suspended matter of sewage and industrial origin are particularly harmful to health.
- Radioactivity: Radioactive pollution occurs due to presence of radionuclides. Radioactive elements are naturally present in earth crust particularly in granite rocks and responsible for background radioactivity. It is very low. Problem arises when radionuclides of high radioactivity are released into water by human activities. Some radioactive elements include plutonium, uranium, thorium, radon, iodine, neptunium, americium, cesium, curium, strontium-90, radium226, cesium-137, phosphorus-32 etc.

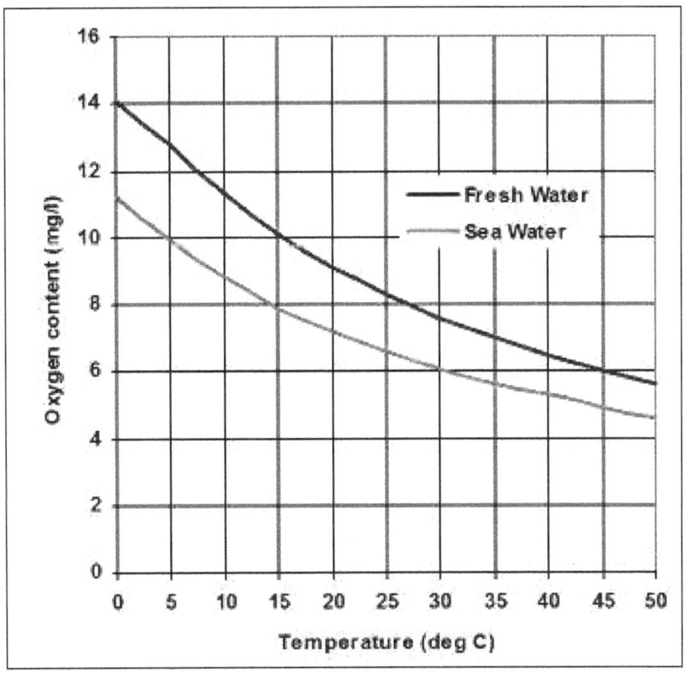

Effect of thermal pollution on dissolved oxygen in water.

- Froth: Froth formation occurs due to presence surfactants like soaps, detergents, fatty acids and saponins etc. these lower the surface tension of water. Sodium lauryl sulfate and Sodium lauryl ether sulfate are common surfactants found in personal care products. Blowing agents like gases (e.g. carbon dioxide), baking powder, isocyanate etc. help in foam formation.

- Thermal pollution: Thermal pollution is increase in temperature of water body due to discharge of heated water or wastewater into the watercourse. Waste heat is mainly discharged into water by nuclear and thermal power plants, and by industries involving water as coolant in industrial manufacturing. Increased water temperature leads to depletion of oxygen and increased respiration.

Physiological Pollutants

Physiological pollution, as said earlier, is caused by presence of some chemicals. Physiological pollutions are of two types- taste and odor.

- Taste: Many chemical compounds are released by industrial waste, which impart unpleasant taste to the water.
- Odor: Odor in water is mainly produced by presence of volatile organics or due to metabolites produced by microorganisms.

Table: Physiological pollutants imparting characteristics taste or odor to water.

	Pollutants	Characteristics
Taste	Synthetic detergents	Mouldy taste
	Phenols	Phenolic taste
	Chlorophenols	Intense, objectionable taste
	Aromatic nitro derivative	Bitter almond taste
Odour	Indole, skatole	Unpleasant fecal odor
	Methyl/dimethyl amines	Fishy odour
	Organo sulfur (methyl/ethyl mercaptans)	Putrid smell
	Hydrogen sulfide	Rotten egg smell
	Organo phosphorus	Fishy smell
	Aldehydes, cyano derivatives	Bitter almond smell
	By-products of microorganism metabolism	Various smells like fishy to grassy

Biological Pollutants

The biological pollution occurs due to introduction and growth of micro and macro organisms in water body, which adversely affect the quality of water, human health and ecosystem. Various biological pollutants include bacteria, algae, weeds, viruses, protozoa and worms. Table shows the various biological pollutants found in water.

Table: Some common biological pollutants.

Biological pollutants	Examples
Bacteria	E. coli, Salmonella, S. typhi, Vibrio cholera, Clostridium botulinum etc.
Algae	Various macro and micro algae, some produces toxins.
Viruses	Hepatitis A virus, Poliovirus etc.
Protozoa	Entamoeba histolytica, Naegleria fowleri, Cryptosporidium parvum, Cyclospora cayetanensis, Giardia lamblia etc.
Worms	Round worms, tape worms, flukes etc.
Weeds	Hydrilla, potamogeton, ceratophyllum etc.

Groundwater Pollution

Groundwater is found underground in soil or under rock structures called aquifers. About 25% of the planet's overall fresh water supply is groundwater, while surface water stored in rivers, lakes and soil moisture accounts for less than 1%. Groundwater and surface water are often closely interconnected, so when surface water becomes polluted, groundwater can become contaminated as well. Humans often use aquifers as a means to obtain drinking water, and build wells to access it. When this water becomes polluted it is called groundwater pollution.

Groundwater pollution is often caused by pesticide contamination from the soil; this can infect our drinking water and cause huge problems. Spilling gasoline, oil, paint thinners, etc. onto the ground can contaminate groundwater. Groundwater pollution is a low-risk ecological problem but high-risk health problem. Contaminated water in the aquifer will slowly flow, creating a plume of contaminated water. Contaminants in groundwater are not diluted or dispersed because this water moves usually less than 0.3 meter per day, polluted groundwater has very slow cleansing rate due to the low dissolved oxygen and low bacterial count. It can take hundreds of years to cleanse degradable wastes and non-degradable wastes are there permanently.

Factors Affecting Ground Water Pollution

Point and non-point sources of ground water pollution.

The extent of ground water pollution depends on the following factors:

- Rainfall pattern.
- Depth of water table.
- Distance from the source of contamination.
- Soil properties such as texture, structure and filtration rate.

Some Major Incidences of Water Pollution

Minamata Episode

In Minamata city of Japan, disastrous effects of mercury poisoning, was observed in 1956. Chisso cooperation, a chemical manufacturing company of Minamata, discharged huge amount of pollutants including mercury into the Minamata bay, which causes the disease known as Minamata disease. The symptoms of this disease include numbness in hands and feet, general muscle weakness, lack of coordination in muscles (ataxia), damage to vision, hearing or speech, paralysis, coma and death in extreme cases. This disease claimed thousands of death of humans and other animals. The researchers took 2 to 3 years to find out the reasons of Minamata disease, since the mercury released by chemical company was inorganic and had relatively low toxicity. But a high level of organic form of mercury (methyl mercury) was reported in fish and shellfish of Minamata bay.

Actually inorganic mercury was enzymatically converted into methyl mercury by methane synthesizing anaerobic bacteria. Acidic medium facilitates this conversion. Methyl mercury is very toxic, soluble in water and has ability to bio-accumulate through food chain. It enters the food chain through phytoplankton and concentrates thousands of times in fishes on higher trophic level. Consumption of such fish causes disease in man and other animals. This incidence of Minamata is the best example of how one type of pollutant can be converted into other types by environmental factors.

Itai Itai Disease

This disease, due to cadmium poisoning, was first reported in 1912 in Toyama, Japan. Cadmium was released into the rivers as a result of mining activities in Toyama region and continued for years. The water of Jinzu river and its tributaries was used as drinking water, fishing and other general purposes but mainly in rice fields. Rice absorbs heavy metal, Cadmium. On consumption of contaminated rice the cadmium got accumulated in peoples and manifested its ill effect. Cadmium poisoning causes softening of bones and kidney failure. Symptom of cadmium poisoning that is weakness of bones results in spinal and leg pain. Painful screams of victim "Ouch-ouch" is the literal meaning of itai itai.

Exxon Valdez Oil Spill

It was one of the largest oil spills. It occurred on 24 March 1989, in the gulf of Prince William Sound, Alaska by oil tanker Exxon valdez. From it about 11 million gallons of crude oil was spilled which heavily polluted about 200 miles of area. The spilled oil caused the death of more than a lakh of seabirds, thousands of sea otters and harbor seals, hundreds of bald eagle and very high number of solmon and herring.

Torrey Canyon Oil Spill

It occurred off the south west coast of United Kingdom on 18 March, 1967. The Torrey canyon oil tanker struck the rocks near Isles of Scilly. The pounding and breaking of oil tanker spilled 25-36 million gallon of oil and it is worst spill in UK history. It contaminated hundreds of km of French and Cornish coast. It claimed life of 15000 seabirds and huge number of other marine organisms. Heavy detergents used to mitigate the problem also caused the damage.

Air Pollution

Air pollution is the number one environmental health risk worldwide, according to the World Health Organization (WHO). It is attributed to the premature death of around 7 million people around the world, and the OECD estimated the cost of air pollution-related health care at $ 21 billion in 2015, projecting that this cost could reach $ 176 billion in 2060. On top of this, a huge percentage of people around the world are breathing bad air.

Sources of Air Pollution

When we try to study the sources of Air pollution, we usually enlist a series of activities and interactions that create these pollutants. There are two types of sources that we will take a look, namely Natural sources and Man-made sources.

Natural Sources

Natural sources of pollution include dust carried by the wind from locations with very little or no green cover, gases released from the body processes of living beings (Carbon dioxide from humans during respiration, Methane from cattle during digestion, Oxygen from plants during Photosynthesis). Smoke from the combustion of various inflammable objects, volcanic eruptions, etc. along with the emission of polluted gases also makes it to the list of natural sources of pollution.

Man-made Sources

While looking at the man-made contributions towards air pollution, it can be further divided into: Outdoor pollution sources and Indoor pollution sources.

Outdoor Pollution Sources

The major outdoor pollution sources include power generation, vehicles, agriculture/waste incineration, industry and building heating systems. Smoke features as a prominent component. The smoke emitted from various forms of combustion, like in biomass, factories, vehicles, furnaces, etc.

Waste dumped in landfills generates methane, which is harmful in several ways. The reactions of certain gases and chemicals also form harmful fumes that can be dangerous to the well-being of living creatures.

Indoor Pollution Sources

In low-and middle-income countries, mostly burning fuels such as dung, coal and wood in inefficient stoves or open hearths produces a variety of health-damaging pollutants. These include carbon monoxide, methane, particulate matter (PM), polyaromatic hydrocarbons (PAH) and volatile organic compounds (VOC). Even burning kerosene in simple wick lamps also produces significant emissions of fine particles and other pollutants. Exposure to smoke from cooking fires causes 3.8 million premature deaths each year.

Various Causes of Air pollution

1. The Burning of Fossil Fuels: Sulfur dioxide emitted from the combustion of fossil fuels like coal, petroleum for energy in power plants, and other factory combustibles is one the major cause of air pollution. Billions of vehicles run on roads are powered by gasoline and diesel engines that burn petroleum for releasing energy. Petroleum is made up of hydrocarbons, and engines don't burn them cleanly.

As a result, pollutants such as PM, nitric oxide and NO_2 (together referred to as NO_x), carbon monoxide, organic compounds, and lead emit from vehicles including trucks, jeeps, cars, trains, airplanes, causing a high level of pollution. These modes of transportation form part of our daily basic needs, so we rely on them heavily.

But, their overuse is killing our environment as dangerous gases are polluting the atmosphere. Carbon Monoxide caused by improper or incomplete combustion and generally emitted from vehicles is another major pollutant along with Nitrogen Oxides, that is produced from both natural and man-made processes. As per the World Health Organization (WHO), exposure to outdoor air pollution contributes to as much as 0.6 to 1.4 percent of the burden of disease and 4.2 million deaths every year.

2. Agricultural Activities: Ammonia is a very common byproduct of agriculture-related activities and is one of the most hazardous gases in the atmosphere. The use of insecticides, pesticides, and fertilizers in agricultural activities has grown quite a lot. They emit harmful chemicals into the air and can also cause water pollution. Farmers also set the field and old crops on fire in order to keep them clean for the next round of sowing. The burning to clean fields is said to cause pollution by releasing harmful gases in the air.

3. Waste in Landfills: Landfills are land areas in which waste is deposited or buried. These deposited or buried wastes generate methane. Methane is a major greenhouse gas that is highly flammable and very hazardous. E-waste is another grave concern involving a lot of unscientific dismantlings such as chemical leaching, burning wires and others.

4. Exhaust from Factories and Industries: Manufacturing industries release a large amount of carbon monoxide, hydrocarbons, organic compounds, and chemicals into the air, thereby depleting the quality of air. Manufacturing industries can be found at every corner of the earth, and there is no area that has not been affected by it. Petroleum refineries also release hydrocarbons and various other chemicals that pollute the air and also cause land pollution.

5. Mining Operations: Mining is a process wherein minerals below the earth are extracted using large equipment. During the process, dust and chemicals are released in the air causing massive air pollution. This is one of the reasons which are responsible for the deteriorating health conditions of workers and nearby residents.

6. Indoor Air Pollution: Household cleaning products, painting supplies emit toxic chemicals in the air and cause air pollution. Have you ever noticed that once you paint the walls of your house, it creates some sort of smell that makes it literally impossible for you to breathe? Suspended particulate matter popular by its acronym SPM, is another cause of pollution. Referring to the particles afloat in the air, SPM is usually caused by dust, combustion, etc. Around seven million premature deaths caused every year due to the combined effects of ambient (outdoor) and household air pollution cause, WHO says.

7. Natural Events: There are certain natural events such as volcanoes, forest fires, and dust storms, which are nature-born and cause air pollution.

Disastrous Effects of Air pollution

- Respiratory and Heart Problems: The effects of air pollution are alarming. They are known to create several respiratory and heart conditions like asthma, chronic bronchitis, emphysema, heart attacks and strokes along with cancer, among other threats to the body. Several million are known to have died due to the direct or indirect effects of Air pollution.
- Child Health Problems: Air pollution is detrimental to your health even before you take your first breath. Exposure to high air pollution levels during pregnancy causes miscarriages as well as premature birth, autism, asthma and spectrum disorder in young children.

  It also has the potential to damage early brain development in a child and cause pneumonia that kills almost a million children below 5 years. Children are at a greater risk of short term respiratory infections and pulmonary diseases in areas exposed to air pollutants.
- Global Warming: Another direct effect is the immediate alterations that the world is witnessing due to global warming. With increased temperatures worldwide, an increase in sea levels and melting of ice from colder regions and icebergs, displacement, and loss of habitat have already signaled an impending disaster if actions for preservation and normalization aren't undertaken soon.
- Acid Rain: Harmful gases like nitrogen oxides and sulfur oxides are released into the atmosphere during the burning of fossil fuels. When it rains, the water droplets combine with these air pollutants, becomes acidic and then falls on the ground in the form of acid rain. Acid rain can cause great damage to humans, animals, and crops.
- Eutrophication: Eutrophication is a condition where a high amount of nitrogen present in some pollutants gets developed on the sea surface and turns itself into algae and adversely affects fish, plants, and animal species. The green-colored algae that are present on lakes and ponds are due to the presence of this chemical only.
- Effect on Wildlife: Just like humans, animals also face some devastating effects of air

pollution. Toxic chemicals present in the air can force wildlife species to move to a new place and change their habitat. The toxic pollutants deposit over the surface of the water and can also affect sea animals.

- Depletion of the Ozone Layer: Ozone exists in the Earth's stratosphere and is responsible for protecting humans from harmful ultraviolet (UV) rays. Earth's ozone layer is depleting due to the presence of chlorofluorocarbons, hydrochlorofluorocarbons in the atmosphere. As the ozone layer becomes thin, it will emit harmful rays back on earth and can cause skin and eye-related problems. UV rays also have the capability to affect crops.

Impressive Solutions to Air Pollution

- Use the Public Mode of Transportation: Encourage people to use more and more public modes of transportation to reduce pollution. Also, try to make use of carpooling. If you and your colleagues come from the same locality and have the same timings, you can explore this option to save energy and money.
- Better Household Practices: Discard fireplaces and/or wooden stoves used for heating homes. Use gas logs in place of wood. Also, eliminate the use of gas-powered lawn and gardening equipment. Avoid setting fire to garbage, dry leaves, or other materials in your yard, and lighting bonfires in the open. Try to mulch or compost your yard waste. Use cleaning products and paints that are environmentally friendly.
- Conserve Energy: Switch off fans and lights when you are going out. A large number of fossil fuels are burnt to produce electricity. You can save the environment from degradation by reducing the number of fossil fuels to be burned.
- Understand the Concept of Reduce, Reuse and Recycle: Do not throw away items that are of no use to you. Instead, reuse them for some other purpose. For example, you can use old jars to store cereals or pulses.
- Emphasis on Clean Energy Resources: Use of Clean energy technologies like solar, wind and geothermal is on the rise these days. Governments of various countries have been providing grants to consumers who are interested in installing solar panels for their homes. Undoubtedly, this can go a long way to curb air pollution.
- Use Energy-Efficient Devices: CFL lights consume less electricity than their counterparts. They live longer, consume less electricity, lead to lower electricity bills, and also help you to reduce pollution by consuming less energy.

Several attempts are being made worldwide on personal, industrial, and governmental levels to curb the intensity at which air pollution is rising and regain a balance as far as the proportions of the foundation gases are concerned.

This is a direct attempt at slacking Global warming. We are seeing a series of innovations and experiments aimed at creating alternate and unconventional options to reduce pollutants. Air pollution is one of the larger mirrors of man's follies, and a challenge we need to overcome to see a better tomorrow.

Soil Pollution

Soil, a nature's wonderful gift shaped the lives of all the living beings, directly or indirectly. Soil is the upper layer of earth that c is made of mineral and organic constituents. It is produced due to complex processes of modifications of the solid crust. Soil is a living medium in which plants grow and it provides habitat to various organisms. Soil supports cultivation of vegetation to fulfill human and animal food requirements. Linked to primary food chain of ecosystem polluted soil can result into virtual destruction of the lower life forms on chemical effects. Lower food chain pyramid level foreign organic chemicals become more telescoped for each consumers of the food chain. The contaminated soil affects the ecological balance of any system. Most plants are incapable to adaption when the chemistry of the soil changes so radically in a short period of time. Soil erosion becomes another problem when fungi and bacteria begin to decline that bind the soil together. The fertility of the soil slowly recedes, making the land incompatible for agriculture and other local vegetation to sustain. Soil pollution is caused by human activities, especially unhygienic habits, wrong agricultural practices and solid and liquid waste disposal.

Past land use, current activities on the site and nearness to pollution sources greatly impinge the soil properties. Substances such as pesticides, fertilizers and other amendments have knowingly added to soil by humans. Accidental spills and leakage of chemicals which are used for commercial or industrial purposes have also been an additive source of contamination.

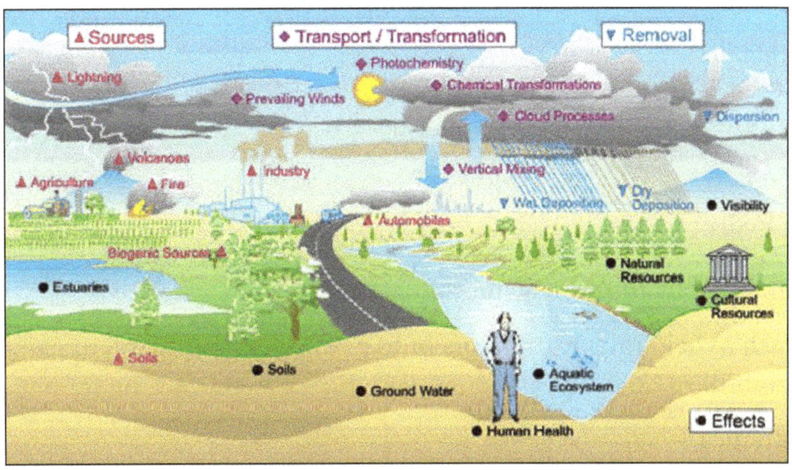

Sources of soil pollution.

Some pollutants are moved through the air and deposited as dust or by precipitation. Penetration of synthetic pesticides and insecticides deteriorate the soil quality. Atmospheric pollution is also responsible for soil pollution to a great extent. Entry of xenobiotic components and changes in the natural environment can lead to soil pollution. These harmful chemicals settle on the soil due to the effect of gravity and pollute the soil. Hydrocarbons, polyaromatic carbons, petroleum, lead and other heavy metals are the most common type of pollutants.

Agent Orange was one of the weapons used during the Vietnam War. It had adverse impact on human and environment which included skin cancer in soldiers, and even infertility. Also due to its impact as a defoliant, it was later banned. DDT (dichlorodiphenyltrichloroethane) and Gammaxene were first widely used as insecticides at the end of World War II but insects soon developed resistance to these chemicals and owing to their synthetic origin, these chemicals persisted in the environment for long. They entered the food chain and disrupted calcium metabolism in birds leading to thinning of eggshells. Many higher organisms became endangered due to accumulation of these chemicals in their bodies. Falcons, eagles and brown pelicans were a few of them. Though DDT has been banned in many countries but still countries like USA have been exporting this hazardous chemical to other developing countries. The global land deterioration rate has been increased perilously in recent years and prone to hike more in further decades.

Types of Soil Pollution

There are many different ways which may cause soil pollution:

- Landfill seepage.
- Release of industrial waste into the soil.
- Leakage of contaminated water into the soil.
- Parting of underground storage tanks.
- Excess use of pesticides, herbicides or fertilizer.
- Discharge of Solid waste.

The most common chemicals involved in soil pollution are:

- Heavy metals.
- Pesticides.
- Solvents.
- Petroleum hydrocarbons.

Soil can be contaminated by three basic mechanisms:

- Rainfall, such as acid rain, rain falling onto a sanitary landfill, oil or chemical waste spilled onto the ground.
- Human activities.
- Physico-chemical alterations, which allow polluting substances to move within or between soil layers.

Soil pollution may be caused by any substance that poses harm to living organisms. Such substances deteriorate soil quality and even disturb the natural composition of soil leading to soil erosion. The Sources and effects of contaminants are helpful in distinguishing various types of soil pollution. Based on this, three types of pollution have been identified – Agricultural, Industrial and Urban pollution.

Agricultural Sources

- Agricultural practices devote to soil pollution profoundly. Although fertilizers enhanced crop yield but they greatly cause pollution and effects soil quality. The massive application of nitrogenous and phosphate fertilizers induced acidifying effect to the soil. There are various processes which decide the fate of fertilizers in the soil are: plant and animal uptake, adsorption, exchange in the soil, leaching or infiltration of ground water, gaseous losses to the atmosphere and surface loss in the solid form by runoff and erosion.

- Similarly, Pesticides spray also cause soil pollution. Not only this, these harmful chemicals penetrate deep inside the soil and even may contaminate the ground water aquifers causing diseases life blue baby syndrome, etc.
- Surface runoff due to rain and irrigation also contaminate the nearby water streams and aquatic life.

Sources of agricultural pollution.

Industrial Sources

- Improper waste disposal contaminates the soil due to addition of harmful and unwanted chemicals thus affecting plant and animal species.
- These chemicals are also found to contaminate local water streams.
- About 90% of oil pollution is caused by industrial waste products.
- Toxic fumes that arise from the regulated landfills contain chemicals that return back to earth as acid rain and damage the soil profile.

Industrial pollution

Urban Activities as a Source of Soil Pollution

- Human activities have direct as well as indirect impacts on soil quality.
- Improper drainage and increase run-off contaminates the nearby land areas or streams.
- Improper solid and liquid waste disposal deposits a number of chemicals and pollutants onto the soil which further seep into groundwater and contaminate the local water system.
- Due to excessive deposition of waste, there is decrease in bacterial population in the soil that deems the soil unfit.
- Decomposition by bacteria generates methane gas contributing to global warming which is a greenhouse gas.

Impacts and causes of urbanization on soil pollution.

Causes of Soil Pollution

Industrial Activity

From the last century industrial activity has been the biggest grantor to the problem, principally since the extent of mining and manufacturing has expanded. Most of the industries are dependent on extracting minerals from the soil. The contaminated i.e. coal, iron ore or it's by products are not disposed off in a manner that can be considered safe. As a result, the industrial waste dawdled in the soil surface for a long time and makes it inadmissible for use.

Agricultural Activities

Due to modern use of pesticides and fertilizers, the chemical utilization has been gone up staggeringly. These fertilizers are made up of chemicals that are not synthesized in nature and so cannot be broken down by it. Due to this the seepage occurs and gets mixed with the ground water and slowly dwindled the fertility of the soil. Also the composition of the soil gets damaged by the chemicals and erosion becomes easier by air and water. Plants also absorb these chemicals and when they disintegrate becomes part of land and hence causes soil pollution. Besides human sources, trace metals can also be found in the parent material from which the soil is developed. However, the amount of these elements may be increased due to several human activities i.e. applications of chemicals, sewage sludge, farm slurries, etc.

Long-term and extensive use of land for agriculture with frequent application of agrichemicals is one of the major causes of trace metal, such as copper, nickel, zinc and cadmium, accumulation in soil. Accumulation of Cu in agricultural soil is a consequence of the century – old practice of using copper-sulphate (Bordeaux mixture) and other copper containing fungicides to control vine downy mildew. Widespread distribution of Cd and its high mobility makes it a potential contaminant in a wide range of natural environments. Phosphatic fertilizers are one of the most ubiquitous sources of Cd contamination in agricultural soils throughout the world. Soils may get contaminated due to addition of heavy metals in soil by application of increased doses of the fertilizers, pesticides or agricultural chemicals. A number of factors required to decide whether the said inputs will be toxic or not and to what extent. These factors are:

- Specific chemical and physical trace metal characteristics.
- Soil type.
- Land use.
- Geo-morphological characteristics within the soil type.
- Exposure to emission sources.

The processes which affect the mobile transformation and toxicity of metals in soil are of great importance in the rhizosphere which is actually the zone where soil root development takes place. Due to this we need to understand the behavior of trace elements in the soil, how they are built up in the soil and their uptake by the plants. Metals are chemically very susceptible in the environment which may result in the mobility and bioavailability to living organisms. Trace metals should be in soluble state or bound to mobile particles in order to be translocated to the soil profile by water.

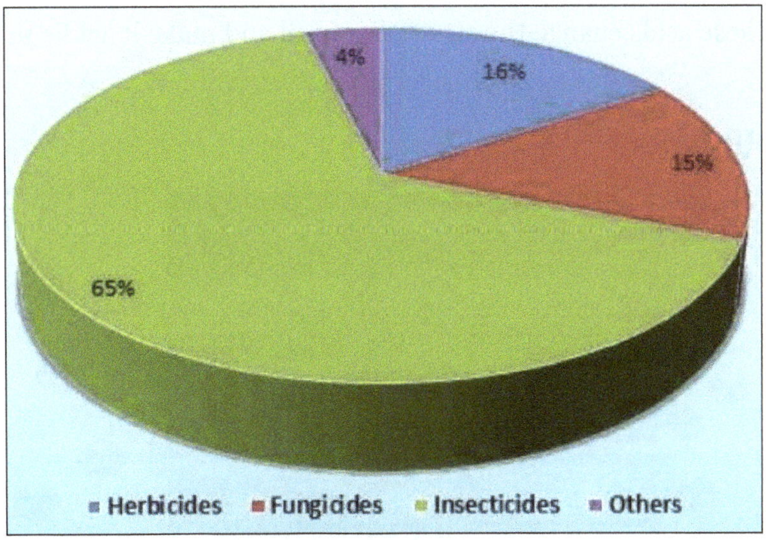

Pesticide use pattern in India.

Complexes are formed due to the combination of metals and particles of organic matter present in the topsoil which can be translocated angularly along the profile depth. So many health problems like vomiting, abnormalities, nausea and gastrointestinal occurs due to long term deposition of heavy metals. For e.g. Arsenic from smelter emissions and pesticides when used in the soil remains for hundreds of years in it. Pesticides in agriculture clings chemicals in the environment for a long time which can lead to the soil poor quality and also affects the beneficial organisms.

Waste Disposal

Finally, a growing cause for concern is how to dispose of our waste. While industrial waste is sure to cause contamination, there is another way in which we are adding to the pollution. Every human produces a certain amount of personal waste products by way of urine and feces. Though most of it founds its fate in the sewers but a large amount is dumped on the landfills in the form of diapers. Even the sewage ends up stacking up into the landfills where the biological waste pollutes the soil and water.

Accidental Oil Spills

Oil spills can happen during storage and transportation of chemicals. This has been observed at most of the fuel stations. The chemicals present in the fuel deteriorate the quality of soil and make them unsuitable for cultivation. These chemicals can enter into the groundwater through soil and also make the water undrinkable.

Acid Rain

Sulphur dioxide (SO_2) and oxides of nitrogen and ozone to some extent are the primary causes of acid rain. These pollutants originate from human activities such as combustion of burnable waste, fossil fuels in thermal power plants and automobiles. These constituents interact with reactants present in the atmosphere and result into acid deposition. When air pollutants combine with the rainwater, it forms acid rain. Along with the wet deposition there are also dry depositions of acids, which can be transformed into salts in the soil and cause the same environmental damage, as do the wet deposits. These acids change the structure of soil and make it acidic thus deeming it unfit for cultivation.

Radioactive Pollutants

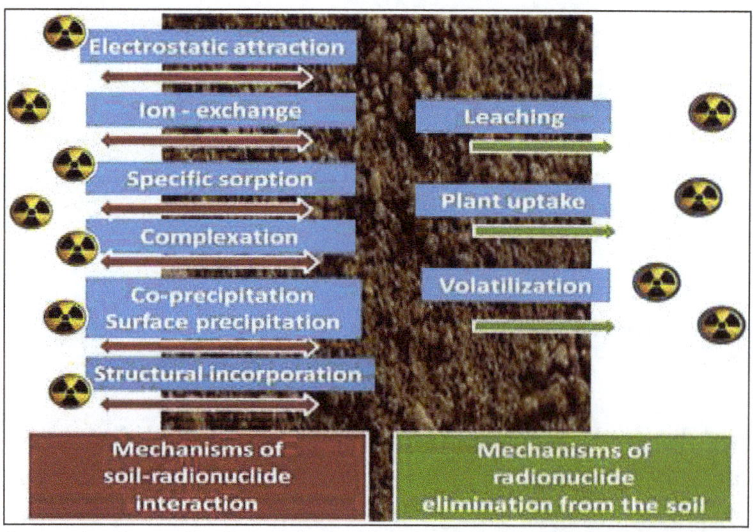

The mechanisms of radionuclide binding and elimination from soil matrix.

Explosions and wastes from nuclear testing laboratories and industries generate radioactive dust which penetrates into the soil thus contaminating it. On the basis of toxicity radioactive waste is classified as high, intermediate, or low level toxic. A waste can remain radioactive from seconds to minutes, or even for millions of years depending on the radionuclide contained in it. The radionuclide distribution in the soil includes the source term and the release conditions, transport and dispersion mechanisms, and the properties of the ecosystem. In the soil the nature of radionuclide retention and the affinity to certain soil constituents depends upon its chemical form and the reactivity.

The charges on the surface of the soil particles, and the chemical properties help to establish covalent bonds the interactions between radionuclide and the soil. Most of the sand and silt fraction of the soil mainly quarts and feldspar are derived from the parent rock. Due to the relatively low specific surface area, their role in contaminant interaction is the smallest, and the attachment occurs through reversible sorption. Minerals such as clay result from weathering processes including physical, chemical, and biological processes. With small particle size and large specific surface area clay are the carriers of permanent surface charge which make them important matrices for contaminant retention. Consequently, clay minerals, Fe, Al-oxides are abundant in amorphous form, and organic matter, containing functional groups, such as OH^- and $COOH^-$ undergo a variety of interactions with contaminants.

- Uranium, Radium, Thorium are radio nucleoids, Potassium (K-40) and Carbon (C-14) isotopes are normally found in soil, rock, water and air.
- Explosion of hydrogen weapons and cosmic radiations include neutron, proton reactions by which Nitrogen (N-15) produces C-14 which participates in Carbon metabolism of plants and then into animals and human beings.
- Radio nuclides such as Strontium-90, Iodine-129, Cesium-137 and isotopes of Iron which are highly toxic are present in radioactive waste. In bones and tissues of humans and animals Strontium is found its way to get deposited.
- The harmful effect of radioactive material depends upon their half-life for e.g. nuclear reactors produce waste containing Ruthenium-106, Iodine-131, Barium-140, Cesium-144 and Lanthanum-140 along with primary nuclides Sr-90 with a half-life 28 years and Cs-137 with a half-life 30 years hence more the half-life more will be the toxic effect.
- Sr-90 and Cs-137 deposited on the soil through rain water where they are tightly held with the soil particles by electrostatic forces. Once deposited on the soil all the radio nuclides emit gamma radiations.

Climate Change

It is another important factor which likely to affect all the degradation threats, and hence soil properties. Soils on the edge of stability are likely to be effected by the changes in the seasons like to drier summers, low rainfall soil suffer irreversible shrinkage upon drying on the other hand in rainy season increase the risk of flooding and salinization in low-lying coastal soils, one major effect of flooding is reduction of total microbial biomass. Climate change will affect important soil functions while effecting soil properties.

Effects of Soil Pollution

Soil system is very dynamic, complex and important ecological factors. Every plant depends on it for their nutrient and water supply. Past land use practices and current human activities are responsible for soil pollution. To have maximum from land pesticides, fertilizers and other amendments to soils have been added by farmers. Accidental spills and leaks of chemicals used for commercial or industrial purposes are other sources of soil contamination. Another factor like acid rain increases the exchange between hydrogen ion and nutrient cations like potassium (K^+), magnesium (Mg^{2+}) and calcium (Ca^{2+}) in the soil leading to acidification of soil. These cations are rapidly leached out in soil solution along with sulphate from acid input which results in nutrient deficiency and altogether disturbing the nutrient cycle in the affected soils. The loss of soil fertility results into decrease in the growth of plants.

The number of hydrogen and aluminum ions in solution increases and the available nutrients decrease due to acidity. This will disturb food chain and microorganism along with the whole ecosystem. If we take the example of oil then In assessing the environmental situation relating to oil contamination, it is important not only to determine the spread of pollution, but also the depth of its penetration. There have been cases of large-scale hydrocarbon accumulation in soils in the territories of oil-gas production enterprises, refineries, transport pipelines and accidental spills. Oil products are characterized by high migration capacity and enter into deep horizons of soil. Most oils have certain content of metals. Concentrated in asphaltenes and tars heavy metals can cause deleterious effect on the soil. Among metals like Pb, Cd, Hg, Zn and Cu can be very noxious to the ecosystem and human beings. Enzyme system function get disturb as majority of heavy metals have strong affinity for sulfur forming bonds with sulfur groups in enzyme. They can easily enter a food chain and reach toxic levels due to their high persistence. These may ultimately kill many species of fish, birds and mammals.

Effect on Human Health

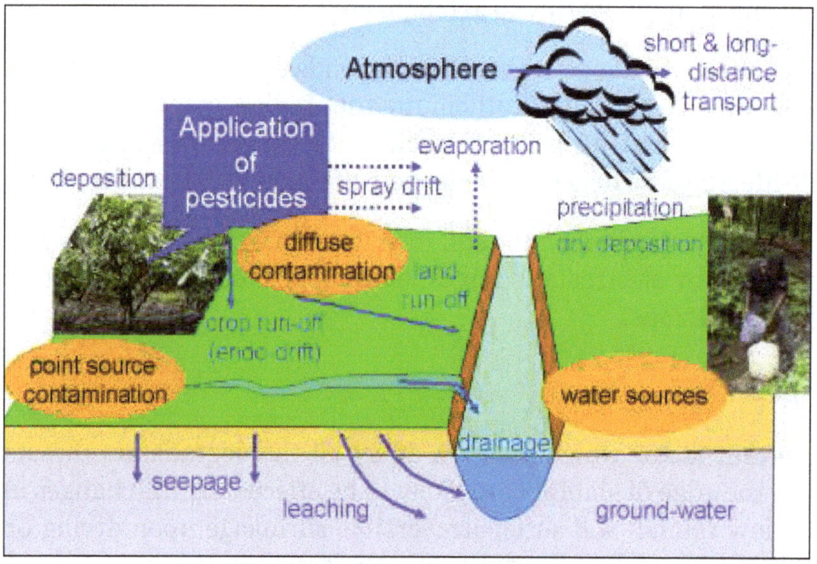

The effect of contaminants on soil pollution.

Since we are dependent on the soil system for food and other reasons, soil health is of utmost importance. People can get soil contaminants from the animals raised for food from contaminated soil, and people by consuming animal products such as meat, eggs and milk may be exposed to these contaminants. Various diseases are also transfer to us by the crops and plants grown on polluted soil for e.g. salmonella being a food borne disease, one of the common route of this bacteria in vegetable is through soil. Plants retain those toxins and when they die, it decompose the toxic material back into the soil and in due course the soil will become unusable. Many bacterial infestations spread through contact with polluted soil like, Leptospirosis, Actinomycosis, Anthrax, Botulism etc. It is through contaminated soil that many helminthes like Ascaris lumbricoides, Necator americanus, Ancylostoma duodenale etc. enters human beings and some other examples like Tetanus bacteria and some Hanta virus are transmitted to human through infected rodent excreta which may be in soil. Long term exposure to such soil can affect the genetic make-up of the body, causing congenital illnesses and chronic health problems that cannot be cured easily.

Effect on Growth of Plants

Elevated levels of soil contaminants can negatively affect plant vigor, animal health, microbial processes, and overall soil health and hence the ecological balance of any system gets affected. Most plants are unable to adapt when the chemistry of the soil changes so drastically in a short period of time. For example, legume plants are able to fix nitrogen in the soil through a symbiotic relationship with Rhizobium bacteria in their root nodules of crops (including beans, lentils, peas, and peanuts) and are often used to replenish nitrogen levels in depleted soils. However zinc contamination can disrupt the nitrogen fixation process as bacteria are sensitive to it and hence key nutrient nitrogen for plant growth may then no longer be available to the plant or to the rest of the system.

Moreover fungi and bacteria found to help in the soil aggregation begin to decline, which leads to an additional problem of soil erosion. Soil loses its fertility gradually making it unfit for cultivation not able to support most types of vegetation just like deserts. Some contaminants may change plants' metabolic processes and reduce yields or cause visible damage to crops like oil pollution. Oiled shoots of crops may wilt and die off stomatal blockage thereby inhibiting respiration photosynthesis and transpiration. Germination, growth performance and yield of these crops are hindered by oil spillage. Oil and oil products are characterized by high migration ability and penetrate into deep horizons of soil. There have been cases of large-scale hydrocarbon accumulation in soils in the territories of oil-gas production enterprises, refineries, transport pipelines and accidental spills.

Decreased Soil Fertility

The toxic chemicals present in the soil can decrease soil fertility and therefore decreases land production. The addition of toxic chemicals which impact chemical properties of soil that regulates life processes is known as Chemical degradation process. These toxic chemicals also known to affect other processes of soil also like nutrient availability, nutrient uptake and nutrient element mobility. Contaminated soil is then used for horticulture purpose. Fruits and vegetables thus produced deficient in nutrients and may have poisonous substances that definitely going to cause serious health problems in people consuming them.

Agricultural inputs or contaminants also affect many soil chemical properties functions, mainly pH. From the atmosphere or the use of fertilizers, agrochemicals, manures or waste materials on land the deposition and accumulation of metals and metalloids in topsoil takes places which, under acidifying conditions if released in toxic concentrations, can severely impair plant growth and food quality. Pesticides reaching the soil and causing degradation of some aspect of soil properties cannot be underestimated and such possibilities become high when pesticides are applied at high rates over many years and most of them remain persistent in soil varying degrees.

Pesticides enter soils from spray drift during foliage treatment, wash off from treated foliage, release from granulates or from treated seeds in soils. Pesticides in soils have been identified to have major effect on soil microbes. Significant irreversible changes in soil microbial populations are being caused by Pesticides by affecting soil micro flora and micro fauna and harm soil fertility. Petroleum hydrocarbon also affects soil fertility, thus affecting the physicochemical properties of the soil such as temperature, structure, nutrient status and pH. The oil obstruct proper soil aeration as oil film on the soil surface acts as a physical barrier between air and the soil, it 'sterilize' the soil and prevent crop growth and yield for a long period of time. The diversity and activity of soil microbes play a vital role in recycling of plant nutrients, maintenance of soil structure and detoxification of noxious chemicals.

Crude oil spillage is one of the factors that affect soil micro flora activities in the soil. The fungi population abundance may decrease with time since, by the depletion in nitrogen due to biodegradation as it is being used up during biodegradation of petroleum compounds. This reduces the amount of nitrogen available to the flora hence their population would decline. In the oil contaminated soils hydrocarbon-oxidizing microorganisms are absent and biological self-purification system of soil gets hampered. This results into increase of accumulation of exchangeable bases in the oil polluted soils in comparison to Exchangeable cations (Ca^{2+}, K^+, Mg^{2+}, Na^+) which further affects the ionic stability of the soil. The absence of nitrogen, phosphorus and other biogenic elements and diffusion of atmospheric oxygen all that are necessary for microbial activity are blocked due to hard saturation of soils with heavy oil fractions–asphaltenes and tars leads to complete suppression of microbiological degradation in contaminated soil. To curtail this problem, liming, fertilization, enhanced ploughing and harrowing are recommended. Further the farmers are advised to carry out remediation processes on oil polluted soils and not to cultivate untill.

Changes in Soil Structure

Soil structure usually gets alter after death of many soil organisms (e.g. earthworms) in the soil apart from this; it could also force other predators to move to other places in search of food. Degradation of soil physical properties, particularly the decrease in porosity from compaction and the destabilization of structure from loss of SOM, affects a range of soil functions. Water storage and flood regulation functions are degraded with a loss in porosity. Poor aeration of the soil which results in root stress might also be a factor. If we take the e.g. of petroleum products which are very toxic to living organisms in soils which indirectly control the chemical and biochemical activities in soils for plant.

As a result of crude oil pollution which is denser than water may reduce and restrict permeability, soil physical properties such as pore spaces might be clogged which reduces soil aeration, infiltration of water into the soil, increased bulk density, depletion of soil nutrients and inhibition of enzymatic and soil microbial activities of the soil which may affect plant growth. Another factor i.e.

climate change also have influence on soil structure (type, spatial arrangement and stability of soil aggregates) is a more complex process. The most important direct impact especially during heavy rains, thunderstorms and even 'rain bombs' is the aggregate-destructing role of raindrops, surface runoff and filtrating water, the increasing hazard, frequency and intensity of which are characteristic features of climate change. The indirect influences are caused by changes in the vegetation pattern and land use practices. The activities of microorganisms, especially beneficial microorganisms are so important to nutrient release to crops/vegetation grown on a particular soil hence the need for avoidance of pollution and contamination of soil is vital.

Effect of Chemicals on Soil

Salinization is a process of chemical soil degradation, which greatly reduces soil productivity. Kavvadias defines salinization as the accumulation of water-soluble salts (including sodium, potassium, magnesium and calcium, sulphate, carbonate and bicarbonate) on or near the surface of the soil. Salinization involves the accumulation of different salts, but the increased content of exchangeable sodium (Na^+) in a soil resulting to a completely unproductive soil is referred to as sodification. Salinization is a common problem in arid and semi-arid regions where the rate of water loss from the soil through evaporation and transpiration is higher than the amount of rainfall received. Acidification is the change in the chemical composition of the soil, which may trigger the circulation of toxic metals Acidification impacts negatively on the soil ecosystem thereby causing damage to plants. It also results in the alteration of soil water chemistry. Soil acidification results from pH decline or from acid deposition. The phenomenon of acid deposition arises from the deposition of emissions from vehicles such as SO_2, power stations, other industrial processes and natural bio-geochemical cycles onto the soil surface mainly via rainfall and dry deposition Acidified soils hinder the availability of some mineral elements in the soil either by reacting to produce forms that become bound to the soil particles or form complexes or the elements are leached further down the soil structure.

Under the influences of man and land use soil is always responding to changes in environmental factors. Some changes in the soil will be of short interval and reversible, others will be an everlasting feature of soil. Due to the presence of soil pollutants soil productivity may be completely lost. Soil pollutants have an undesirable effect on the physical chemical and biological properties of the soil and reduce its efficiency. We should avoid soil that has been contaminated to grow food, because the chemicals will enter into the food chain and harm people.

To avoid the contamination proper methods should be adopted for waste disposal. Industrial wastes should be treated physically, chemically and biologically until they are less hazardous to be disposed off. Soil acidity and alkalinity is being caused by acidic and alkaline wastes which should be first neutralized; Biodegradable waste should be allowed to degrade under controlled conditions before being disposed off. Time to use bio-fertilizers and manures which will reduce chemical fertilizer and pesticide use. Biological methods of pest control can also reduce the use of pesticides and thereby minimize soil pollution. It is important to note that degradation processes do not necessarily have irreversible effects on soil functions. Land management practices can be revised to reduce or even remove the degradation threat to improve soil properties and functions.

References

- Air-pollution: plumelabs.com, Retrieved 02, March 2020
- Causes-effects-solutions-of-air-pollution: conserve-energy-future.com, Retrieved 17, August 2020

3

Environmental Waste and its Management

Environmental waste encompasses different categories of waste such as hazardous waste, radioactive waste, municipal solid waste, e-waste and biomedical waste. Its management involves the processes which are responsible for handling and disposing off this waste safely. The topics elaborated in this chapter will help in gaining a better perspective about these types of environmental waste and their management.

Waste is defined as any unwanted or unusable substance that is discarded after primary use. Wastes are normally generated as a result of human and animal activities. Urbanization and rapid advancements in industrialization has led to an increase in the production and consumption processes resulting in the generation of wastes from various sectors that include agricultural, commercial, domestic, industrial, institutional, social and from community activities. Over time, these wastes accumulate and can have real impacts on the health and the environment. Waste management is intended to reduce adverse effects of waste on health, the environment or aesthetics. So to carry out efficient management of wastes, knowledge about the source of wastes, its types and classification must be known.

Classification of Wastes

Wastes are commonly classified based on the physical state as solid wastes, liquid wastes and gaseous wastes.

- Solid wastes are any discarded or abandoned materials that can be solid, liquid, semi-solid or containerized gaseous material discarded by the human society. These include urban wastes, agricultural wastes, biomedical wastes and radioactive wastes. The term refuse is also used for solid wastes. Examples of solid wastes include waste tires, septage, scrap metal, latex paints, furniture and toys, garbage, appliances and vehicles, oil and anti-freeze, empty aerosol cans, paint cans and compressed gas cylinders, construction and demolition debris, asbestos, plastics, styrofoam containers, bottles etc.
- Liquid wastes: Liquid wastes can be defined as liquids/fluids that are generated from washing, flushing or manufacturing processes of the industries. They are also called as sewage. The most common practice of disposing liquid waste is to discharge it in ground or rivers and other water bodies without treatment. Examples: Domestic washings, chemicals, oils, waste water from ponds, Wastewater from manufacturing industries, manure, waste oil, fats, oils or grease (FOG), used oil, and hazardous household liquids.
- Gaseous wastes: It is a waste product released in the form of gases from automobiles, factories, industries, burning of fossil fuels etc. and get mixed in the atmosphere. These gases include carbon monoxide, carbon dioxide, sulphur dioxide, nitrogen dioxide, ozone and methane etc.

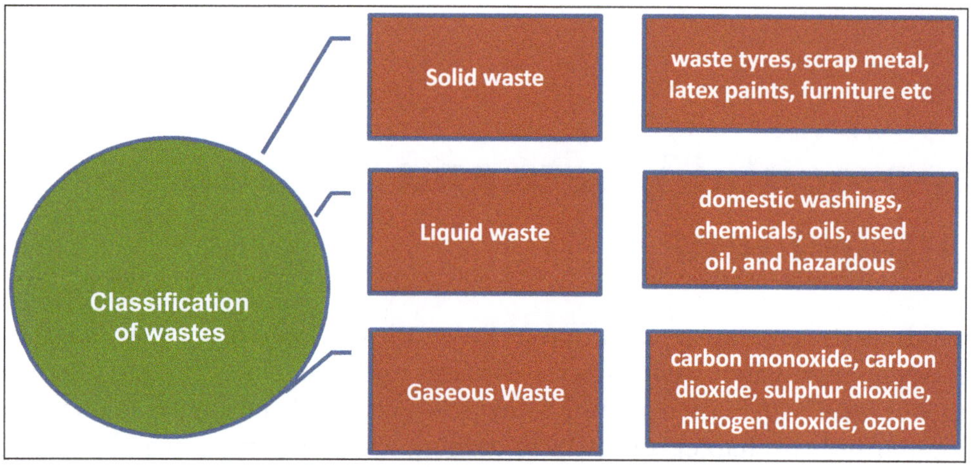

Classification of waste based on their physical state.

Hazardous Waste

Wastes are classified as hazardous if they exhibit one or more of ignitability, corrosivity, reactivity, or toxicity. Hazardous wastes are defined as any waste or combination of wastes which pose a substantial present or potential hazard to human health or living organisms because such wastes are non-degradable or persistent in nature or because they can be biologically magnified, or because they can be lethal, or because they may otherwise cause or tend to cause detrimental cumulative effects.

The management of hazardous wastes has become a specialized discipline because of the complex nature of the problem and the solutions available to humanity. The mismanagement examples of hazardous wastes causing disastrous human and environmental consequences are numerous. The management process is based on the definition and classification of the different wastes, and their toxic effects on human and taking in consideration the application of risk management to control human health and environmental impacts of hazardous waste. Hazardous waste management, therefore, deals with minimizing harmful effects on humans and environment by applying special techniques of handling, storage, transportation, treatment, and disposal of hazardous wastes.

Hazardous Characteristics

A useful listing of hazardous characteristics is that provided by the United Nations as part of recommendations relating to the transport of dangerous goods.

Table: Hazardous characteristics: Extracted from UN listing.

UN class number	Hazardous characteristic
1	Explosive
3, 4	Flammable
5	Oxidizing
6	Poisonous/infectious

7	Radioactive
8	Corrosive
9	Toxic (delayed or chronic)/eco-toxic

Industrial Wastes

Waste generated from industrial sources can have non-hazardous and hazardous components, with non-hazardous waste usually representing the greater part of the volume. The hazardous component of this waste is relatively small in volume. This type of waste was identified as hazardous waste when proceeds toxicity test, corrosively test, ignitability test, and some special character test. As a hazardous pollutant, it may impose serious impacts on surrounding environment and such impacts should be quantitatively examined to assess the influence on human health.

Household Waste

Households generate small quantities of hazardous wastes such as oil-based paints, paint thinners, wood preservatives, pesticides, insecticides, household cleaners, used motor oil, antifreeze, and batteries. It has been estimated that household hazardous waste in industrialized countries such as the United States accounts for a total of about 0.5% (by weight) of all waste generated at home, while in most developing countries, the percentage probably is even lower.

Biomedical Waste

There are some of hazardous medical and dental wastes that, when disposed improperly, could cause harm to the environment. It also presents an occupational health hazards to the healthcare personnel who handle these wastes at the point of generation and those involved with their management, that is, segregation, storage, transport, treatment, and disposal. Healthcare waste that is capable of producing injury or disease including many sorts of hazardous wastes such as:

- Infectious waste: Which contain pathogens namely bacteria, viruses, fungi, or parasites in concentrations sufficient to cause disease in susceptible hosts. Cultures and stocks of infectious agents from laboratory work; tissues and dressing generated from autopsies, surgeries, and treatment of infected patients and animals; materials or equipment in contact with blood and infected body fluids.
- Pathological waste: Including tissue, organs, body parts, human fetuses, and animal carcasses, blood and body fluids.
- Sharps: It comprise syringes, needles, scalpels, saws, infusion sets, knives, blades, broken glass, or other items that can cause cut or puncture wounds.
- Pharmaceutical waste: It covers expired, unused, spilt, and contaminated pharmaceutical products, drugs, vaccines, and sera that are no longer required and need to be disposed of in appropriate manner.
- Genotoxic waste: This type combines cytostatic drugs, vomit, urine, or feces from the patients treated with cytotoxic drugs, chemicals, and radioactive materials. Genotoxic waste has mutagenic, teratogenic, and carcinogenic properties.

- Chemical waste: Discarded solid, liquid, or gaseous chemicals should be considered as hazardous if it is toxic, corrosive, inflammable, or reactive.
- Waste with high content of heavy metals: Mercury (thermometers, blood pressure gauges, amalgam), cadmium (discarded batteries), and lead (reinforced wood panels for radiation proofing in radiology department) generated from hospitals could be represented as a subcategory of hazardous chemical waste.
- Radioactive waste: The use of radioisotopes in vitro analysis of body tissues and fluids, in vivo organ imaging, tumor localization, and treatment and various clinical studies involving certain radionuclides need to be specially managed in a centralized treatment facility for radioactive wastes.

Nuclear applications have been rapidly developed recently, and several nuclear power plants started to work throughout the world. The potential impact of released radioactive contaminants into the environment has received growing attention due to nuclear accidents. Contamination of soil and water by radionuclides due to natural processes, global fallout from nuclear weapon testing, discharges from nuclear installations, disposal of nuclear waste, and occasional nuclear accidents poses serious problems to biological systems. Radioactive waste includes a variety of radionuclides and occurs in a variety of physical and chemical forms. It can be generally classified as low-/intermediate-level radioactive waste and high-level radioactive waste.

Radioactive waste, arising from civilian nuclear activities as well as from weapon activities, poses a potential problem for handling and saving the environment for coming generations. Radioactive waste includes a variety of radionuclides and occurs in a variety of physical and chemical forms. It can be generally classified as low/intermediate-level radioactive waste and high-level radioactive waste. Nuclear research establishments include, for example, waste containing different organic components, toxic or chemically aggressive constituents, radionuclides with specific properties (high mobility, high chemical activity, volatile elements, etc.), waste difficult for treatment and not appropriate for direct immobilization (e.g., spent organic ion exchange resins and spent liquid scintillation cocktails). For such waste, application of conventional treatment and conditioning options may not be efficient and appropriate in terms of economy, safety, and performance characteristics. In many cases, such wastes are stored awaiting an appropriate treatment and conditioning solution.

The primary sources of radioactive wastes in a country without nuclear fuel cycle activities are nuclear research, production of radioisotopes, application of radioisotopes, and decontamination and decommission of nuclear installations.

Treatment and Disposal

Hazardous Waste Reduction and Minimization

Due to the high potential for public health and environmental damage, some hazardous wastes require special control procedures. Management of these wastes means awareness and control over them from the time of generation through their transportation, temporary storage, treatment and disposal. Effective identification and labelling of wastes by the generators is essential to the effective operation of any manifest-based system. In addition, management of hazardous waste means more than careful disposal. It implies consideration of alternate methods and schemes, both institutional and technical, to hazardous waste reduction. The following measures are commonly adopted as part of Hazardous waste reduction and minimization measures:

- Elimination: The primary priority is to find ways to eliminate uses of hazardous substances altogether. Elimination might be achieved by changing manufacturing processes or by substituting products that can satisfy the same need without creating hazardous wastes (an example would be the substitution of concrete posts in place of, creosote-treated wooden posts).
- Reduction: The next priority is to reduce the amount generated. Again manufacturing processes can be modified to yield substantial waste reduction. For example, changes in the chemical reaction conditions can minimize production of by-product hazardous substances. In some instances, potentially hazardous catalysts, can be replaced by catalysts that are non-hazardous or that can be recycled rather than discarded.
- Recycling: The third strategy is to recycle hazardous substances such as solvents and acids to maximize their use before treatment and disposal becomes necessary. Many industrial processes that use solvents are equipped for solvent recycle for reasons of both economics and pollution control. A number of operations are used in solvent purification. Entrained solids are removed by settling, filtration, or centrifugation. Drying agents may be used to remove water. Fractional distillation is used to separate solvent from impurities, water and other solvents.
- Finally, hazardous substances can be treated to reduce their volume and toxicity.

Eliminating, reducing, and recycling hazardous substances reduces treatment and disposal costs and often the generators can find it in their own economic best interest to actively seek out these source reduction opportunities as a way to reduce the overall cost of doing business.

Hazardous Waste Treatment Technologies

The selection of a treatment process for a waste stream depends on among other factors the nature of the waste, desired characteristics of the output stream. Even with after vigorous hazardous waste reduction program, there will still be large quantities of hazardous wastes that will require treatment and disposal. The treatment technologies have been categorized as physical, chemical, biological, thermal, or stabilization/fixation.

- Physical treatment processes include gravity separation, phase change systems, such as air and steam stripping of volatiles from liquid wastes, and various filtering operations, including carbon adsorption.
- Chemical treatment transforms waste into less hazardous substances using such techniques as pH neutralization, oxidation or reduction, and precipitation. Biological treatment uses microorganisms to degrade organic compounds in the waste stream.
- Thermal destruction processes include incineration, which is increasingly becoming a preferred option for the treatment of hazardous wastes, and pyrolysis, which is the chemical decomposition of waste, is brought about by heating the material in the absence of oxygen.

- Stabilization techniques involve removal of excess of water from a waste and solidifying the remainder either by mixing it with a stabilizing agent such as Portland cement, or vitrifying it to a glassy substance.

Most treatment measures have both physical and chemical aspects. The appropriate treatment technology for the hazardous wastes depends on the nature of the wastes. The type of physical treatment to be applied to wastes depends strongly upon the physical properties of the material treated, including the state of matter, solubility in water and organic solvents, density, volatility, boiling point and melting point.

Physical Treatment Methods

Adsorption

Adsorption on activated carbon occurs when a molecule is brought up to its surface and held there by physical and/or chemical forces. This process is reversible, thus allowing activated carbon to be regenerated and reused by proper application of heat and steam, or solvent. The factors that relate to adsorption capacity are:

- Greater surface area produces greater adsorption capacity (Eg: Activated carbon has large surface area (500-1500 m^2/g)].
- Adsorptivity increases as the solubility of the solute (in solvent) decreases. Thus, for hydrocarbons, adsorption increases with molecular weight.
- For solutes with ionizable groups, maximum adsorption will be achieved at a pH corresponding to minimum ionization.
- Adsorption capacity decreases with increasing temperature.

One additional point to be noted is that biological activity usually takes place in a carbon bed. If the concentration of the adsorbed species is high enough and the material is biodegradable and nontoxic to the bacteria, then biological activity may significantly increase the effective removal capacity. Removal through adsorption by activated carbon has been applied to non-aqueous waste stream such as petroleum fraction, syrups, vegetable oils, and pharmaceutical preparations. Color removal is the most common objective in such cases. Current waste treatment applications are limited to aqueous solutions.

Resin Adsorption

Waste treatment by resin involves two basic steps: (1) contacting the liquid waste stream with resin and allowing the resin to adsorb the solutes from the solution; and (2) subsequently regenerating the resins by removing the adsorbed chemicals, by simply washing with proper solvent.

The adsorption of a nonpolar molecule on to a hydrophobic resin (e.g., styrene divinyl-benzene-based resin) results primarily from the effect of Vander Waal's forces. In other cases, other type of interactions such as dipole-dipole interaction and hydrogen bonding are important. In a few cases ion exchange mechanism may be involved. For the removal of organic dye wastes from water, two different resins were employed: In this case the waste stream is first contacted with a normal polymeric adsorbent and then with an ion exchange resin.

Sedimentation

Sedimentation is a physical process whereby particles suspended in a liquid settle by means of gravity. The fundamental elements of most sedimentation processes are:

- A basin or container of sufficient size to maintain the liquid to be treated in a relatively quiescent state for a specified period of time.
- A means of directing the liquid to be treated into the above basin in a manner conducive to settling.
- A means of physically removing the settled particles from the liquid.

Sedimentation can be either a batch or a continuous process. Continuous processes are by far the most common, particularly when large volumes of liquid are to be treated. This technique has been widely used in the removal of heavy metals from iron and steel industry wastewater; removal of fluoride from aluminium production wastewater; and removal of heavy metals from wastewater from copper smelting and from metal finishing industry and wastewater stream from organic chemicals.

Electrodialysis

The electrodialysis involves the separation of an aqueous stream (more concentrated in electrolyte than the original) and a depleted stream. Success of the process depends on special synthetic membranes, usually based on ion exchange resins, which are permeable only to a single type of ion. Cation exchange membranes permit passage only of positive ions, under the influence of electric field, while anion exchange membranes permit passage only of negatively charged ions.

The feed water is passed through compartments formed by the spaces between alternating cation-permeable and anion permeable membranes held in a stack. At each end of the stack is an electrode having the same area as the membranes. A dc potential applied across the stack causes the positive and negative ions to migrate in opposite directions. This technique has been used for desalination to produce potable water from brackish well water. In food industry electrodialysis is used for desalting whey. The chemical industry uses this technique for enriching or depleting solutions, and for removing mineral constituents from product streams.

Reverse Osmosis

This technique which is most widely used consists of a membrane permeable to solvent but impermeable to most dissolved species, both organic and inorganic. These devices use pressure to force the contaminated water against the semipermeable membrane. The membrane acts as a filter, allowing the water to be pushed through the pores, but restricting the passage of larger molecules that are to be removed.

Cellulose acetate membranes were used in the past, but nowadays polysulphones and polyamides are increasingly popular for use at high pH values. Because of the susceptibility of the membranes to chemical attack and fouling, and the susceptibility of the flow system to plugging and erosion, it is common to pre-process the feed water to remove oxidising materials. The reverse osmosis technique has been widely used for desalination of sea or brackish water. It has also been successfully used in the treatment of electroplating rinse waters, not only to meet effluent discharge standards, but also to recover concentrated metal salt solutions for reuse. It has also been used for treatment of waste stream from paper and food processing industries.

Solvent Extraction

Solvent extraction is the separation of the constituents of a liquid solution by contact with another immiscible liquid. If the substances comprising the original solution distribute themselves differently between the two liquid phases, a certain degree of separation will result and this may be enhanced by the use of multiple contacts. The major application of solvent extraction to waste treatment has been in the removal of phenol from by-product water produced in coal coking, petroleum refining, and chemical synthesis that involve phenol.

The use of supercritical fluids (SCFs) most commonly CO_2 as extraction solvent, has been one of the more promising approaches to solvent extraction. SCFs are fluids existing at or above the lowest temperature at which condensation may occur. Above the critical temperature certain fluids exhibit characteristics that enhance their solvent properties. Organic materials, which are only slightly soluble in particular solvents at room temperature, become completely miscible with the solvent when under supercritical conditions. The excellent solvent properties result from the rapid mass transfer ability and the very low density that characterizes an SCF. Major advantages of SCFs are short residence times with no char formation. Some of the important application of these SCFs, has been in the extraction of organohalide pesticide from soil, extraction of oil from emulsions used in aluminium and steel processing, and regeneration of spent activated carbon. Supercritical ethane has been used to purify waste oils contaminated with PCBs, metals and water.

Distillation

Distillation is expensive and energy intensive and can probably be justified only in cases where valuable product recovery is feasible (e.g., solvent recovery). This technique has only limited application in the treatment of dilute aqueous hazardous wastes.

Evaporation

Evaporation process is used for the treatment of hazardous waste such as radioactive liquids and sludges and concentrating of plating and paint solvent waste among many other applications. It is capable of handling liquids, slurries and sometimes sludges, both organic and inorganic, containing suspended or dissolved solids or dissolved liquids, where one of the components is essentially non-volatile. It can be used to reduce waste volume prior to land fill disposal or incineration. The major disadvantages of evaporation are high capital and operating costs and high energy requirements. This process is more adaptable to waste waters with high concentrations of pollutants.

Filtration

Filtration is well-developed economical process used in the full scale treatment of many industrial waste waters and waste sludges. Energy requirements are relatively low, and operational parameters are well defined. However, it is not a primary treatment process and is often used in conjunction with precipitation, flocculation, and sedimentation to remove these solids.

Flocculation

The various phenomena that occur during flocculation can be grouped in to two sequential mechanisms:

- Chemically induced destabilization of repulsive surface related forces, thus allowing particles to stick together when they touch.
- Chemical bridging and physical enmeshment between the non-repelling particles, allowing for the formation of large particles.

Chemicals used for flocculation include alum, lime, ferric chloride, ferrous sulphate and poly electrolytes. Poly electrolytes consist of long chain, water soluble polymers such as polyacrylamides. They are used either in conjunction with inorganic flocculants, or as primary flocculating agent. The inorganic flocculants such as alum, upon mixing with water, the slightly higher pH of water causes them to hydrolyze to form gelatinous precipitates of aluminium hydroxide. It is partially because of their large surface area, they are able to enmesh small particles, and thereby create larger particles.

Once suspended particles have been flocculated into larger particles, they usually can be removed from the liquid by sedimentation, provided that a sufficient density difference exists between the suspended matter and the liquid.

Chemical Treatment Methods

Chemical Oxidation and Reduction

1. Oxidation reduction is one important route used for the treatment of hazardous wastes. It is a process by which the hazardous wastes are converted into a form that shows reduced or no toxicity, reduced or no mobility and remains inert. The process can be short or medium term technology and uses oxidizing agents for the treatment of hazardous waste. Few examples of oxidizing agents include ozone, hydrogen peroxides, chlorine, chlorine dioxide and others. Some examples of oxidation reduction treatment of hazardous wastes are given below. Chemical redox treatment applied for cyanide wastes released from metal finishing industry is given below. Chlorine in alkali solution is used, which converts the CN to cyanate (less toxic form). More addition of chlorine oxidizes the cyanate to CO_2 and N_2.

$$NaCN + Cl_2 + 2NaOH \rightarrow NaCNO + 2NaCl + H_2O$$

$$2NaCNO + 3Cl_2 + 4NaOH \rightarrow 2CO_2 + N_2 + 6NaCl + 2H_2O$$

2. Hexavalent chromium Cr(VI) is reduced to trivalent chromium Cr(III) by using Sulphur dioxide as reducing agent.

$$3SO_2 + 3H_2O \rightarrow 3H_2SO_3$$

$$2CrO_3 + 3H_2SO_3 \rightarrow Cr_2(SO_4)_3 + 3H_2O$$

3. Iron (II) is oxidize and precipitated as ferric hydroxide.

$$4Fe^{2+} + O_2 + 10H_2O \rightarrow 4Fe(OH)_3 + 8H^+$$

4. Sulphur dioxide is oxidized to sulphuric acid.

$$2SO_2 + O_2 + 2H_2O \rightarrow 2H_2SO_4$$

Ozonolysis

Ozone is used to oxidize a wide variety of contaminants present in waste water and sludge. Ozone is generated by an electrical discharge through dry air or oxygen. Ozone is also used to destruct TCDD and PCBs. TCDD demonstrated that if the dioxins were suspended as an aerosol in combination with CCl_4, degradation of 97% TCDD would be possible. Destruction of polychlorinated phenols and pesticides by ozone can be done in conjunction with UV radiation. In both the cases the key requirements were to concentrate the TCDD in a medium where they were susceptible to attack and provide a free radical for reaction with dioxin molecule.

$$(CH_2O) + 2[O] \rightarrow CO_2 + H_2O$$

$$CH_3CHO + [O] \rightarrow CH_3COOH$$

Acid-base Neutralization

A waste showing a pH less than 2 or more than 12.5 are categorized as corrosive hazardous wastes. They can be neutralized by acid base reactions. Lime is used to neutralize acid wastes. Addition of slaked lime $[Ca(OH)_2]$ to acidic wastes in a continuously stirred chemical reactor results in neutralization of hazardous waste. Lime is commonly preferred due to its economic nature and optimum solubility. Due to limited solubility, excess lime does not increase the pH values. Similar to lime, sulphuric acid and acetic acid is used for neutralizing alkaline wastes. Though sulphuric acid is inexpensive, at certain times acetic acid is preferred as it is nontoxic and biodegradable. Certain alkaline wastes are also neutralized by bubbling gaseous CO_2 forming carbonic acid. CO_2 is easily available and can be collected from the exhaust gas of any combustion process. In addition to this, may other waste treatment processes require prior pH adjustment. The process include oxidation/ reduction, adsorption, wet air oxidation, ion-exchange, stripping and biochemical treatment.

Chemical Precipitation

Precipitation is applicable for mixed wastes that contain both organic and inorganic species. By this method the toxic substances or metals are precipitated using chemicals and removed from the solution by sedimentation or settling or filtration. Alteration of solution pH will minimize the solubilization of heavy metals which can further be precipitated as precipitable hazardous constituent. Slaked lime $[Ca(OH)_2]$ or caustic soda is used for precipitation of the metal ions as metal hydroxides. For example, the following reaction suggests the use of lime to precipitate the metal as hydroxide.

$$M_2 + Ca(OH)_2 \leftrightarrow M(OH)_2 + Ca^{2++}$$

Chromium is precipitated as hydroxide.

$$Cr^{3+} + 3(OH^-) \rightarrow Cr(OH)_3$$

Sodium carbonate is also used to precipitate metals as hydroxides. ($Fe(OH)_3 . XH_2O$), carbonates ($CdCO_3$), basic carbonate salts ($2PbCO_3.Pb(OH)_2$).

Carbonate ion hydrolyses in water to give hydroxide ion.

$$CO_3^{2-} + H_2O \rightarrow HCO_{3-} + OH^-$$

Metals can be precipitated as sulphides from solution. Precipitating the metals as sulphides will facilitate removal of low concentration of metals. Ferrous sulphide is popularly used. Sodium borohydride is another reducing agent that precipitates metal in elemental form.

$$4Cu^{2+} + NaBH_4 + 2H_2O \rightarrow 4Cu + NaBO_2 + 8H^+$$

Hydrolysis

Hydrolysis treatment can be given to those hazardous waste constituents which shows high reactivity with water. Examples of those substances are halides, carbide, hydride, alkoxide, and active metal.

$$SiCl_4 + 2H_2O \rightarrow SiO_2 + 4HCl$$

$$CaC_2 + 2H_2O \rightarrow Ca(OH)_2 + C_2H_2$$

$$NaAlH_4 + 4H_2O \rightarrow 4H_2 + NaOH + Al(OH)_3$$

$$NaOC_2H_5 + H_2O \rightarrow NaOH + C_2H_5OH$$

$$Ca + 2H_2O \rightarrow Ca(OH)_2 + H_2$$

Ion Exchange

Ion exchange is best suited for wastes that have undergone pre-treatment or minimum treatment with other chemical and physical methods. It is suitable for removing inorganic chemical species that are in the dissolved form unlike precipitation, flocculation or sedimentation. Basically it is used as a final polishing step to reduce an inorganic species that could not be reduced to satisfactory levels by preceding treatment processes. Nickel cyanide complex and chromate ions from hazardous waste solutions are removed by anion ion exchangers.

$$2Res\,OH^- + \left[Ni(CN)_4\right]^{2-} \rightarrow (Res^+)_2\left[Ni(CN)_4\right]^{2-} + 2OH^-$$

$$2Res\,OH^- + CrO_4^{2-} \rightarrow (Res^+)_2(CrO_4^{2-}) + 2OH^-$$

Ion exchange resins have also been used in the removal of radionuclides from radioactive wastes.

Thermal Treatment Methods

Thermal destruction of hazardous waste is done by incineration and pyrolysis. Incineration is a process that combusts the wastes at high-temperature (900-1100 °C) in the presence of oxygen. This process converts waste to an inert, less toxic and less bulky material namely, the ash. Ash is a non-combustible inorganic residue. Incineration reduces the volume of waste by 90%. The organic elements such as carbon, hydrogen and oxygen are converted wholly or partially to gaseous form. The principal products of incineration are CO_2, water vapor and ash.

$$C(\text{organic}) + O_2 \rightarrow CO_2 + \text{heat}$$

$$4H(\text{organic}) + O_2 \rightarrow 2H_2O + \text{heat}$$

The heat generated during the incineration process is used for destruction of organochlorine hazardous wastes. The heat cleaves the C-Cl bonds of organochlorine compounds. If the organochlorines are inflammable, then supplementary fuel such as methane or petroleum-liquid is used. Though incineration process comes with many advantages, still its application is limited due to the release of by-products of hazardous waste incineration such as sulphur, nitrogen, halogen and heavy metals (mercury, arsenic, selenium, lead and cadmium). This results in problems of air pollution.

To overcome the air pollution problem, the incinerators need to be fitted with air pollution control equipment. Temperature, Availability of sufficient oxygen and ample residence time are the critical factors governing incineration process. Destruction removal efficiency (DRE) is used to assess the efficiency of hazardous-waste incinerators. DRE can be calculated as the percentage of mass difference of input(feed) and output (stack emission) waste constituents through the incinerator. DRE has to be calculated separately for different compound/constituent. According to RCRA (Resource conservation and Recovery Act) requirement, a minimum DRE of 99.99 % for most of organic compounds and a DRE of 99.9999 % for dioxins and dibenzo furans should be achieved for efficient combustion.

Advantages of Incineration

- Reduces 90% volume of waste.
- Can be scaled to handle large volumes of liquid waste.
- It is the best method for “mixed wastes” and biologically hazardous wastes.
- Does not require large area of land.
- Compact units.

Disadvantages of Incineration

- Capital and operation cost is very high.

- The ash must be disposed of properly.
- The gaseous and particulate products of combustion should be controlled by air pollution control devices.

Wet Air Oxidation

Wet air oxidation is the process by which the dissolved or suspended organic substances are oxidized in aqueous phase at high temperature and pressure of 150-325 °C and 2000 kPa to 20,000 kPa, respectively. Water acts as a catalyzing agent and moderator by removing excess heat in the reaction mixture. Water also acts as an excellent heat transfer medium so that the wet air oxidation process becomes thermally self-sustaining at low organic feed conditions. In wet air oxidation, the waste is pumped into the system through a heat exchanger into the reactor with high-pressure pump and mixed with air from an air compressor. High pressure facilitates dissolving of oxygen in water. Within the reactor, the organic waste is reacted with atmospheric oxygen. Sometimes the oxidation is carried out in the presence of catalysts. Cyanide from electroplating waste solutions are removed by wet air oxidation process. The process can be used for the removal of,

$$2Na^{+} + 2CN^{-} + O_2 + 4H_2O \rightarrow 2Na^{+} + 2HCO_{3^-} + 2NH_3$$

Photolysis

In photolysis, light is used as an agent to cleave the chemical bonds of hazardous waste materials. Initially, the chemical species is excited by light. It is usually a diradical state because diradical possess sufficient energy to undergo chemical reaction. The success or effectiveness of the process depends on the chemical structure of the compound, efficiency of radiation source and medium in which the reaction is carried out. The end product also determines the efficiency of the process. Polychlorinated biphenyls (PCBs) and chlorinated dibenzo-p-dioxins(CDDs) are destroyed by photolysis. The three requirements of photolysis of tetrachloro dibenzo-p-dioxin (TCDD) are 1) Dissolution in a light transmitting film 2) presence of organic hydrogen odor and 3) ultraviolet light. Generally, during photolysis reactions, HO is formed which plays an important role in chain reactions resulting in destruction of the compound.

Biological Treatment of Hazardous Wastes

Biological processes are a simple, cost effective and eco-friendly process. However its application to hazardous waste is slightly limited due to toxic nature of the waste which is harmful to the biological agents. The biodegradability of the waste depends on the physical and chemical properties of the compound. Selection of appropriate organism and optimum conditions can favor degradation of chemical compounds that are considered harmful or biocidal. Eg. Phenol and phenolic compounds can be treated by proper acclimatization of microorganisms to the hazardous waste. Likewise, biorefractory compounds eg. DDT and PCBs are also degraded by bacteria. DDT is degraded by Pseudomonas and PCB by anaerobic bacteria. Optimum pH and oxygen levels can result in better degradation of waste. Both aerobic and anaerobic bacteria are used for hazardous waste biodegradation.

The presence of inorganic substances limit the biodegradation of waste as they inhibit enzymatic activity and further get adsorbed on to the microbial cell coating. Landfill leachate is treated with microorganisms. During the process H_2S is generated which precipitates toxic heavy metal ions as their respective sulphides. The overall degradation of the hypothetical organic compound (CH_2O) can be written as follows:

$$2(CH_2O) \rightarrow CO_2 + H_2O$$

Land Treatment

The application of waste to soil surface in a controlled manner is termed as land treatment of hazardous wastes. This process facilitates containment of waste followed by biological and chemical degradation of organic waste. This type of treatment is different from landfilling. Here, the soil is used to detoxify, immobilize, and degrade all or a portion of the applied waste. Landfills contain the waste and control its migration from the site. Hazardous wastes should not be placed in a land treatment site unless the waste can be made less hazardous or non-hazardous by the reactions occurring in the soil. The waste is added or mixed to the aerobic soil horizon in a controlled manner at appropriate conditions. By this process, the physical, chemical and biological characteristics of the waste are altered. Bacteria, fungi and actinomycetes play a major role in land treatment. Genus Agrobacterium, Anthrobacteri, Bacillus, Flavobacterium, Pseudomonas are some of the bacteria involved in degradation. Wastes from petroleum refining, biodegradable organic chemical wastes including organochlorine compounds are suitable for land treatment. Land treatment should not be applied to waste containing acids, bases, toxic inorganic compounds, salts, heavy metals and excessively soluble volatile and flammable organic compounds.

Land Disposal

After treatment, the hazardous wastes are converted into suitable form for disposal. Immobilization, stabilization, fixation, and solidification are some of the popularly used techniques for preparation of waste for disposal. Land disposal techniques include Landfills, surface impoundments and underground injection wells.

Landfills

In order to protect public health and environment the design of hazardous waste landfills should be adequate. Figure below shows the typical cross section of a hazardous waste landfill.

The liner system, waste emplacement and cover system are the three components of landfills. Liners should be impermeable and avoid the contact between waste and the surrounding environment. Leachate collection system, drains and leak detectors too form an integral part of the landfill system. They ensure that leachate does not leave the disposal site. Hazardous wastes are contained in secured landfills provided with double liner system. The liner system consists of natural (clay), synthetic material (polymers), and a leachate collection. The primary barrier is a synthetic polymeric flexible membrane with a separation layer of thin geotextile or polymeric material (HDPE, LDPE, PVC etc). Secondary barrier is composed of clay, bentonite enhanced soil or geosynthetic clay.

A leachate collection tubes are layered between the waste and the protection layer. It channelizes the leachate to a pumping station where they are removed and taken up for further treatment. An intermediate high permeability drainage layer is incorporated between primary and secondary barrier. This layer is for the collection of leachate and gas. Every barrier is separated by geotextile fabric. The landfill is sloped to avoid infiltration of water and run off associated with it. The leachate that is collected is treated by (1) passing it through a column containing sorbent material such as carbon or flyash. (2) subjected to suitable physical-chemical units such as chemical addition, flocculation, sedimentation, pressure filtration, pH adjustment, and reverse osmosis to remove the dissolved waste. The secondary leachate collection system acts as backup system which will be useful during failure of primary leachate system. Continuous monitoring of ground water from the wells will help us to determine the failure of primary and secondary liner system.

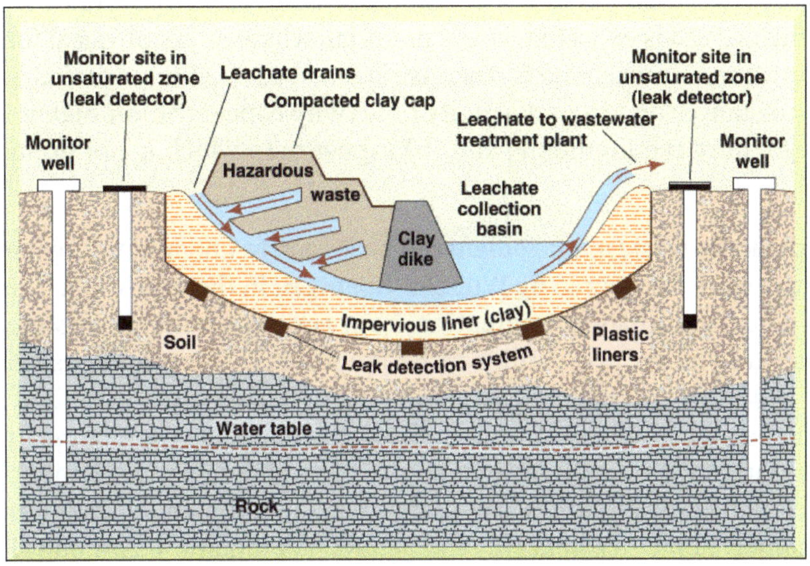

Schematic of hazardous waste landfill.

Waste is loaded inside the landfill after preliminary treatment (i.e) solidification, stabilization etc. The cover or capping is also essential component of the waste landfill. The purpose of the landfill cover is to protect the waste, prevent the entry of rain and surface water into the waste, control the release of landfill gas, to prevent ingress of air and enable the growth of short rooted plant when covered with top soil. Landfills are allowed with sufficient vent points so that if methane is generated, it may be burned off continuously.

Surface Impoundments

A surface impoundment is a man-made excavation, diked area, or natural topographic depression designed to hold an accumulation of liquid wastes. The construction is similar to landfill where the bottom walls function like a liner which is impermeable to liquids. The surface impoundments are also provided with provision for leachate collection. Geological siting and selection of low permeable material like soil and clay are very important as they will arrest or minimize the chances of pollution.

Underground Injection

Underground injection or deep-well disposal is a method by which the hazardous waste materials are injected into underground strata under pressure. The underground strata must be impermeable and separated from the aquifers. During this process, the chance of chemical reactions with waste constituent and mineral is possible. This will result in corrosion, clogging problems. The main concern with underground injection is the potential for contaminating underground drinking water supplies, if the disposal well is not properly cased or if it is damaged.

Radioactive Waste

Radioactive waste is any material that is either radioactive itself or is contaminated by radioactivity, for which no further use is expected. It is a material, whatever its physical form, left over from practices or interventions for which no further use is foreseen: (a) that contains or is contaminated with radioactive substances and has an activity or activity concentration higher than the level for clearance from regulatory requirements, and (b) exposure to which is not excluded from regulatory control.

To understand radioactive waste management it is necessary to understand what are radiations and their sources. X-rays were discovered by Wilhelm Roentgen in the year 1895. Thereafter, instantly Henri Becquerel detected radiation originating from certain uranium salts similarly to X-rays. Marie and his husband Pierre Curie studied radiation from two uranium ores i.e. pitchblende and chalcolite in year 1898, and isolated two supplementary elements that exhibited radiation similar to that of uranium, but substantially stronger. These two elements were named radium and polonium. Atomic age started with the discovery and isolation of radioactive elements i. e. radium and polonium. Curie classified the radiation according to the direction of deflection in a magnetic field into three types namely alpha (α), beta (β), and gamma (γ) radiation. Antoine Henri Becquerel, a physicist was the first person to discover evidence of radioactivity in 1896. The SI unit for radioactivity the becquerel (Bq), is named after him.

In 1905, Ernest Rutherford identified alpha particles emanating from uranium as ionized helium atoms, and in 1932, Sir James Chadwick characterized as neutrons the highly penetrating radiation that results when beryllium is bombarded with alpha particles. Some other subatomic particles were subsequently identified by modern physics. Management of radioactive waste requires understanding of the sources and effects of alpha, beta, gamma, and neutron emissions.

Table: Properties of Ionizing Radiation.

Particle or photon	Mass (amu)	Electric Charge
Alpha ($_2He^4$)	4	+2
Beta (electron)	5.5×10^{-4}	-1
Gamma (X-ray)	˙0	0
Neutron	1	0
Positron (positive electron)	5.5×10^{-4}	+1

A radioactive atom has an unstable nucleus and they moves towards a more stable condition by emitting an alpha or beta particle or neutron; this emission is frequently accompanied by emission of additional energy in the form of gamma radiation. On account of these emissions, the radioactive atom is transformed into either an isotope of the same element (neutron or gamma only emission) or into an isotope of a different element (alpha or beta emission). This transformation is known as radioactive decay, and the emissions are known as ionizing radiation. The rate of radioactive decay can be expressed by a first-order rate equation:

$$dN / dt = -K_b N_t,$$

where,

- N = the number of radioactive nuclei.
- K_b = a factor called the disintegration constant; K_b has the units of time^{-1}.

Integrating over time, we get the classical equation for radioactive decay:

$$N = N_o e^{-K_b t}$$

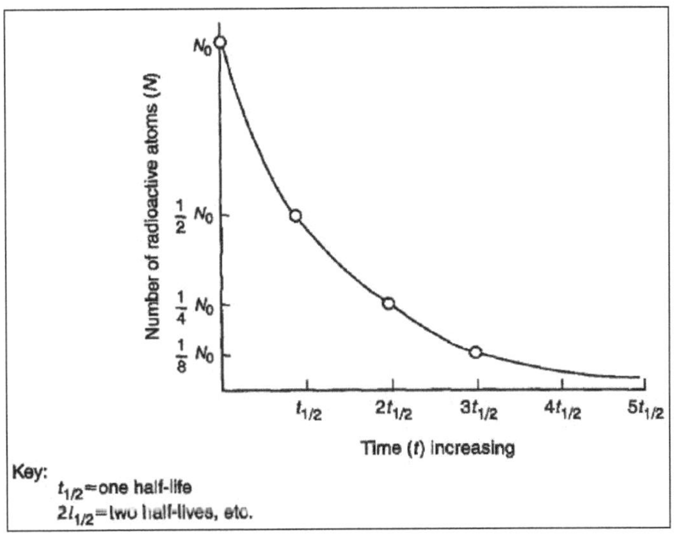

Decay of Radioactive atom with time.

The data points in figure corresponds to this equation. After a specific time period $t = t_{1/2}$, the value of N is equal to one-half of N_o. That is, one-half of the radioactive atoms have decayed (or disintegrated) during each time period $t_{1/2}$. This time period t1/2 is called the radiological half-life, or sometimes simply the half-life. At $t = 2t_{1/2}$, N becomes $1/4N_o$; at $t = 3t_{1/2}$, N becomes $1/8\ N_o$; and so on. N never becomes zero in any finite time period; for every half-life that passes, the number of atoms is halved.

Table: Some Important Radionuclides.

Radionuclide	Type of Radiation	Half-life
Krypton-85	Beta and gamma	10 years

Strontium-90	Beta	29 years
Iodine-131	Beta and gamma	8.3 days
Cesium-137	Beta and gamma	30 years
Tritium (Hydrogen-3)	Beta	12 years
Cobalt-60	Beta and gamma	5 years
Carbon-14	Beta	5770 years
Uranium-235	Alpha	7.1×10^8 years
Uranium-238	Alpha	4.9×10^9 years
Plutonium-239	Alpha	24,600 years

Effects of Radioactive Pollution

The effect of radioactive pollution depends upon:

- Half-life i.e. time of exposure.
- Energy releasing capacity i.e. intensity of radiation.
- Rate of diffusion: This is also referred to as the movement of a substance down a concentration gradient. Diffusion rates increase with temperature, and decrease with increasing pressure, molecular weight, and molecular size.
- Type of ionizing radiation (i.e. its penetration power).
- Various atmospheric and climatic conditions such as wind, temperature, rainfall also determine their effects.

The possible general effects of radioactive wastes are categorized into: Somatic Effect, Genetic Effect and Biomagnification.

- Somatic effect: Affects somatic cells. It appears within individual and disappears with the death of the individual.
- Immediate effects: Anaemia, Reduced immune response, Hemorrhage, skin burn, mouth ulcers, CNS Damage.
- Delayed effects: Eye cataract, Leukemia, Cardiovascular disease, Premature ageing, Reduced life span, reduction of fertility.
- Genetic Effects: The radiation affects the genes of the gamete cells. The changes are not apparent in the individual. The effects are exhibited by offspring and in the subsequent generations. They affect the DNA, RNA replication and chromosome. It causes:
    - Mutation.
    - Chromosomal aberration.

- Chromosomal fragmentation.
- Inhibition of RNA, DNA synthesis.

Table: Effects of radioactive radiation on living beings.

Type of radiation	Effect on the body
α particle	Generally they cannot penetrate the skin. But if their sources are inside the body, they can cause damages to bones or lungs.
β particle	Can penetrate the skin but cannot damage the tissues. They can cause damage to skin and eyes (cataract).
γ radiation	Can easily penetrate the body and pass through it. They cause damage to cell structure.
X rays	Can travel very far and pass through the body tissues except bones. They can cause damage to the cells.

Sources of Radioactive Waste

Radioactive waste is generated from a variety of industries. Any wastes that contain or contaminated with radioactive materials are called as radioactive waste. They can be generated from operation and decommissioning of nuclear facilities, radionuclides using industry, radionuclides containing raw material, medicine and research. Radioactive waste is also generated during the cleanup of sites affected by radioactive residues from various operations or accidents. At certain times, the materials generated as waste may be reused or recycled. Radioactive waste arises from many different activities:

- Operation and decommissioning of nuclear facilities (e.g. nuclear power plants).
- Application of radionuclides in industry, medicine, and research.
- Cleanup of contaminated sites.
- Processing of raw materials containing naturally occurring radionuclides.

Radioactive waste from the nuclear fuel cycle consists of materials with varied range of radioactivity levels. The spent nuclear fuel contains radionuclides like uranium, plutonium, their decay products and many fission products. In many countries, the spent fuel is reprocessed to separate U and Pu. Handling of spent fuel should be done with great care through sophisticated remote handling techniques. Proper shielding should be provided while handling these wastes.

Operational waste includes a variety of lower-level wastes, either in the form of filtering residues (e.g. ion exchange resins) or as contaminated supplies (e.g. clothing) containing fission and activation products. Decommissioning of nuclear power stations gives rise to large volumes of contaminated material, including some highly-activated reactor components as well as structural materials that may be slightly contaminated by activation products and/or fission products.

Wastes generated from other units of the nuclear fuel cycle include milling wastes containing decay products of natural uranium, such as radium and radon and radon daughters etc. Long lifetimes of these materials pose a great problem. Sometimes, the radioactivity is increased when few daughter products are combined. Facilities where the ore is refined, enriched and processed into fuel also result in radioactive waste. In the case of depleted uranium tails, in addition to direct toxicity of the uranium, daughter products (radium and radon) buildup on long time scales also increase the toxicity.

Radionuclides are used extensively in industrial research, agricultural research, geological exploration, constructions for measuring thickness, density or volume of materials. They are also used in smoke detectors; tracers; sealed sources for irradiation, heat/power source.

Classification of Radioactive Waste

In the International Atomic Energy Agency (IAEA) general safety guide 2009, the classification scheme of radioactive waste has been modified with the intention to address the shortcomings identified with the previous scheme and to reflect experience gained in developing, operating and assessing the safety of disposal facilities. The classification scheme developed previously is not completely comprehensive in that it does not cover all types of radioactive waste, nor does it provide a direct linkage with disposal options for all types of radioactive waste. These aspects of the former classification scheme have been deemed limitations on its use and application.

A comprehensive range of waste classes has been defined and general boundary conditions between the classes are provided. More detailed quantitative boundaries that take into account a broader range of parameters may be developed in accordance with national programs and requirements. In cases when there is more than one disposal facility in a State, the quantitative boundaries between the classes for different disposal facilities may differ in accordance with scenarios, geological and technical parameters and other parameters that are relevant to the site specific safety assessment.

A number of elements of the previously set out classification scheme have been retained. A complete range of waste classes has been defined and general boundary conditions between the classes are provided to reflect the experience gained in developing, operating and assessing the safety of disposal facilities. More detailed quantitative boundaries will be prepared according to the national programmes and requirements. In cases when there is more than one disposal facility in a State, the quantitative boundaries between the classes for different disposal facilities may differ in accordance with scenarios, geological and technical parameters and other parameters that are relevant to the site specific safety assessment. Six classes of waste have been derived and used as the basis for the classification scheme:

1. Exempt waste (EW): Waste that meets the criteria for clearance, exemption or exclusion from regulatory control for radiation protection purposes.

2. Very short lived waste (VSLW): Waste that can be stored for decay over a limited period of up to a few years and subsequently cleared from regulatory control according to arrangements approved by the regulatory body, for uncontrolled disposal, use or discharge. Eg. Waste containing primarily radionuclides with very short half-lives often used for research and medical purposes.

3. Very low level waste (VLLW): Waste that does not necessarily meet the criteria of EW, but that does not need a high level of containment and isolation and, therefore, is suitable for disposal in near surface landfill type facilities with limited regulatory control. Such landfill type facilities may also contain other hazardous waste. Typical waste in this class includes soil and rubble with low

levels of activity concentration. Concentrations of longer lived radionuclides in VLLW are generally very limited.

4. Low level waste (LLW): Waste that is above clearance levels, but with limited amounts of long lived radionuclides are categorized as LLW. Such waste requires robust isolation and containment for periods of up to a few hundred years and is suitable for disposal in engineered near surface facilities.

This class covers a very broad range of waste. LLW may include short lived radionuclides at higher levels of activity concentration, and also long lived radionuclides with relatively low levels of activity concentration. Low level Waste (LLW) is generated from hospitals and industry, as well as the nuclear fuel cycle. It comprises paper, rags, tools, clothing, filters etc. which contain small amounts of mostly short-lived radioactivity. It does not require shielding during handling and transport and is suitable for shallow land burial. To reduce its volume, it is often compacted or incinerated before disposal.

LLW includes everything not included in one of the other categories. LLW is not necessarily less radioactive than HLW and may have higher specific activity in becquerel per gram. The distinguishing feature of LLW is that it contains virtually no alpha emitters. To ensure appropriate disposal, the NRC has designated several classes of LLW:

- Class A contains only short-lived radionuclides or extremely low concentrations of longer-lived radionuclides, and must be chemically stable. It is disposed of in designated LLW landfills as long as it is not mixed with hazardous or flammable waste.
- Class B contains higher levels of radioactivity and must be physically stabilized before transportation or disposal. It must not contain free liquid.
- Class C is waste that will not decay to acceptable levels in 100 years and must be isolated for 300 years or more. Power plant LLW is in this category.
- Greater than Class C (GTCC) will not decay to acceptable levels in 300 years. A small fraction of power plant Class C waste is in this category. Some nations like Sweden treat GTCC waste like HLW, and the United States may do the same.

Mixed Low-Level Waste (MLLW) is LLW that contains hazardous waste as defined under the Resource Conservation and Recovery Act (RCRA).

5. Intermediate level waste (ILW): Waste that, because of its content, particularly of long lived radionuclides, requires a greater degree of containment and isolation than that provided by near surface disposal. However, ILW needs no provision, or only limited provision, for heat dissipation during its storage and disposal. Intermediate level Waste (ILW) contains higher amounts of radioactivity and some requires shielding. It typically comprises resins, chemical sludges and metal fuel cladding, as well as contaminated materials from reactor decommissioning. It may be solidified in concrete or bitumen for disposal. ILW may contain long lived radionuclides, in particular, alpha emitting radionuclides that will not decay to a level of activity concentration acceptable for near surface disposal during the time for which institutional controls can be relied upon. Therefore, waste in this class requires disposal at greater depths, of the order of tens of meters to a few hundred meters.

6. High level waste (HLW): Waste with levels of activity high enough to generate significant quantities of heat by the radioactive decay process or waste with large amounts of long lived radionuclides that need to be considered in the design of a disposal facility for such waste. A limit of 400 Bq/g on average (and up to 4000 Bq/g for individual packages) for long lived alpha emitting radionuclides has been adopted in some States. For long lived beta and/or gamma emitting radionuclides, such as 14C, 36Cl, 63Ni, 93Zr, 94Nb, 99Tc and 129I, the allowable average activity concentrations may be considerably higher (up to tens of kilo becquerel per gram) and may be specific to the site and disposal facility. Disposal in deep, stable geological formations usually several hundred meters or more below the surface is the generally recognized option for disposal of HLW. High level Waste (HLW) arises from the use of uranium fuel in a nuclear reactor and nuclear weapons processing. It contains the fission products and transuranic elements generated in the reactor core. It is highly radioactive and hot. HLW accounts for over 95% of the total radioactivity produced in the process of nuclear electricity generation.

Table: Common Categories of Radioactive Waste.

Waste Classes	Typical Characteristics	Disposal Option
1. Exempt Waste (EW)	Activity levels or below clearance levels, which are based on an annual dose to members of the public less than 0.01m Sv.	No radiological restrictions.
2. Low and intermediate level wastes (LILW)	Activity levels above clearance levels and thermal power below about $2KW/m^3$.	Near surface or geological disposal facility.
Short-lived waste (LILW-SL)	Restricted long lived concentrations (limitation of long lived alpha emitting radionuclides to 4000Bq/g in individual waste packages and to an overall average of 400 Bq/g per waste package).	Geological disposal facility.
Long lived waste (LILW-LL)	Long lived radionuclides concentrations exceeding limitations for short lived waste.	
3. High level waste (HLW)	Thermal power above about 2 KW/m^3 and long lived radionuclide concentrations exceeding limitation for short lived waste.	Geological disposal facility.

Management of Radioactive Wastes

Radioactive wastes contaminate air, water, soil and consequently vegetation that eventually affect public health and environment adversely. Since radioactivity is a long term phenomenon, proper and planned disposal of radioactive wastes should be ensured to prevent any negative impact on health and safety of living beings as well as on environmental quality. The principles of radioactive waste management are: (1) protection of human health, (2) protection of the environment, (3) protection beyond national borders, (4) protection of future generations and avoid burden over them (6) national legal framework, (7) control of radioactive waste generation, (8) management interdependencies, and (9) safety of facilities.

The major aim of the environmentalist is to abate the introduction of radioactive materials into the biosphere and environment. Control of the possible direct impact on the human environment is essential but not enough, since radionuclides can be transmitted through water, air, and land pathways for several years and even for generations. Some radioactive waste may be recovered and recycled by future reprocessing, but reprocessing process creates its own radioactive waste stream. The majority of radioactive waste can be treated only by isolating it from the available environment until its radioactivity no longer poses a threat. Isolation requirements vary for different classes of radioactive waste.

Table: Different kind of radioactive waste and its handling.

Type	Origin	Quantity (For 1000 MW station)	Disposal
Low level (LLW)	Waste produced during reactor maintenance.	50 m^3/yr (Including packaging).	Packed in drum and shallow burial.
Intermediate level (ILW)	Waste produced during reactor operation (reactor coolant purification).	50 m^3/yr (Including packaging).	Packed in drum and shallow burial.
High level (HLW)	Radioactive products extracted from spent nuclear fuel (SNF).	3 m^3/yr (Including packaging).	Vitrified packed in capsule and underground repository.

A few radionuclides, mainly those that make up HLW, contain half-lives of tens of thousands of years, or even hundreds of thousands of years. It is difficult to find the technology that solves the problem of ultimate disposal of radioactive waste so long term storage is the only solution till date.

Segregation of Radioactive Waste

The secure, environmentally acceptable and economic management of radioactive waste arising from the nuclear power is an essential component of nuclear power programmes. Noteworthy technical advancement has been achieved and substantial experience of all stages of waste management has been accumulated in many countries. Waste processing, conditioning, storage and transportation have become regular and demonstrably safe activities.

Radioactive wastes should be separated according to their isotopes and placed in separate containers. Tritium (3H) and Carbon-14 (14C) can occupy the same container. Radioactive waste besides segregation based on isotopes, are also separated by physical form. Ten (10) basic physical forms of radioactive waste include: Solid, Glass. Sharps, Liquid, Liquid Scintillation Vials, Biological, Animal Remains, Source Vials, Lead Pigs and Sealed Sources.

Radioactive Solids

It consists of disposable items, radioactive material contaminated lab items such as plastic/rubber gloves, tubing, and syringes; unbroken glassware such as pipettes, beakers, flasks, columns, etc. Specific type of containers is used for collection and storage of solid radioactive waste. Plastic bags are commonly used to collect the solid radioactive waste materials. Yellow radioactive material bags should be avoided for this type of waste expect when they are placed in closeable waste receptacle (for e.g. plastic foot operated trashcan) which remains closed always. The prerequisites for safe handling and disposal of solid radioactive waste include:

- Extreme care while handling radioactive wastes.
- Proper labeling of all bags with the date, month, year, isotope, total activity and department's name.
- Before placing waste into the bags, remove all radioactive and labware labels.
- All the bags should be checked for leaks prior to waste collection. Sometimes, second yellow bag can be used to contain the waste if necessary.
- Take care to segregate sharps to avoid tearing of bags.
- Avoid placing of bags with solid radioactive waste in such places where there are chances of the waste be picked up by housekeeping personnel and disposed off along with municipal waste.
- Waste containers must always remain closed.

Radioactive Glass

It included radioactive material contaminated glassware/other unbroken glass which should be packaged separately from other solid radioactive waste. A strong cardboard box properly labeled and sealed is used for storing the radioactive glass.

Radioactive Sharps

Sharps are defined as anything that could tear the yellow radioactive material bag. Examples: Needles, broken glass, glass pipettes, razor blades, capillary tubes, etc. The container used for collecting such waste must be puncture resistant plastic tubes. These containers once full must be securely capped with orange/red top with complete labelling. The requirements for safe handling and disposal of this type of radioactive waste include:

- Care must be taken to avoid injury during packing of sharps in containers.
- Avoid overfilling of container.
- Ensure that all sharps are dry before placing into container.
- When full, securely cap tube with orange/red top.

Radioactive Liquid

The radioactive liquid waste can be further divided into:

- Aqueous: Aquous liquids are those water based based liquids with a pH between 5.0-9.0, such as saline and buffer solutions that arise from contaminated laboratory glassware washings; weak acids or bases free of biological, pathogenic, or infectious materials.

- Organic: Laboratory solvents such as alcohols, aldehydes, ketones, and organic acids used for various purpose. This category does not include scintillation fluids.
- Other liquids: Example contaminated pump oil.

Specially designated containers termed as carboys are used for disposing radioactive liquids. These carboys should be filled upto 80% of its volume. The prerequisites for safe handling and disposal of this type of radioactive waste include:

- In the laboratory the carboys should be provided with double containment as a precautionary measure against leakage. This will control carboy failure, necessary because pouring is usually accompanied by drips, dribbles, and seeping. Carboys should be placed in a tray or pan that will contain the liquid in the event of a spill of the carboy. At a minimum, plastic backed absorbent paper shall be placed under all liquid waste carboys.
- Do not use glass containers for storage of radioactive liquid waste. If plastic-incompatible contaminated organic solvents are required to be kept in glass containers, the bottle must be doubled contained.
- The first three rinses of the emptied containers must also be placed in the radioactive liquid waste container. No radioactive liquid is to be poured down the sink.
- Avoid mixing of liquid waste types in the carboys (e.g. organic with aqueous).
- Pipettes and other such items should not be placed in the carboys.
- All biological material in the carboys must be properly deactivated. Carboys should be kept free of contamination as far as possible.

Radioactive Liquid Scintillation Vials

Glass or plastic vials containing organic or aqueous based liquid scintillation fluid are included under this category. This type of waste is disposed of in the cardboard trays and placed in a yellow radioactive material bag. At certain cases, these wastes are double bagged in a yellow radioactive material bag for disposal Eg. glass vials are double bagged in yellow radioactive material bags and placed in a cardboard box. Absorbent material must be placed in the cardboard box to absorb any leakage from the vials. Empty vials are to be disposed of as dry and semi-solid radioactive waste. The requirements for safe handling and disposal of this type of radioactive waste are:

- Assure that all tops are fixed tightly on the vials.
- Avoid empting used vials into plastic bags. Leaking or seeping scintillation fluid will dissolve plastic.
- Label each box of vials containing aqueous scintillation cocktail and segregate them from organic cocktail vials.
- If the vials require reusability then pour the liquid waste into a carboy along with the first rinse of the vial. Because most scintillation cocktail solutions are volatile and will contain radioactive substances. The procedures of pouring out the solution should be conducted in a ventilated fume hood.

Radioactive Biological Waste

Radioactive waste containing biological, pathogenic, or infectious material and the equipment used to handle such material are categorized under radioactive biological waste. Examples: By-product animal waste (i.e. serum, blood, excreta), contaminated capillary tubes and other equipment contaminated with animal fluids, radioactive material labeled culture media. Yellow radioactive materials bags labeled with biological waste stickers or red biological bags labeled with radioactive material stickers are the container used for disposal. The prerequisites for safe handling and disposal of this type of radioactive waste include:

- Liquids must be absorbed into another material such as paper towels, sponges, gauze, etc. prior to placing into bags.
- Pathogenic and infectious waste must be sterilized by autoclaving or chemical treatment. If autoclaved, the autoclave must be checked for radioactive contamination after use.

Radioactive Animal Remains

Examples: Radioactive animal carcasses, animal bedding, and by-product animal waste with the carcasses (i.e. viscera, serum, blood, excreta, tissue, etc.) and other animal tissue containing radioactive materials. Yellow Radioactive Materials bags are used as containers for disposing these wastes. The requirements for safe handling and disposal of this type of radioactive waste are:

- Animal remains containing radioactive material in any quantity are subject to handling according to the guidelines. Avoid dumping of these wastes along with municipal wastes.
- A tag showing: The date, radioisotope, total activity, and the laboratory shall be tied to the bag. Bags not labeled cannot legally be disposed of, so an investigation will be performed to identify the generator of the unlabeled waste. If the bag is placed in the freezer by personnel from your lab, the same information must appear on the tag.
- Liquids surrounding carcasses must be absorbed onto another material (e.g. paper towels, sponges, gauze, etc.) prior to placing into yellow bags.

Source Vials Examples

The original vials with radioactive materials shipped from the supplier are categorized under this waste. This includes full, partially full, and empty vials. Source vials must be separated from the dry and semi-solid waste stream and placed in cardboard box for disposal.

Lead Pigs

Examples: Original lead and lead impregnated shielding containers surrounding source vials are included in this category. Lead is a hazardous waste and must be disposed of accordingly. Avoid placing of these items in municipal waste. Cardboard shipping box for used disposal.

All Sealed Sources

Examples: Calibration sources, check sources, quenched standard sets, electron capture gas chromatograph detectors, etc. Sealed sources must be separated from the solid waste and placed in cardboard box for disposal. During disposal, check for broken or crushed sources and handle these damaged sources with extreme care. All sources must be disposed of properly by a private contractor, even if decayed. A final survey or leak test on all sources should be done prior disposal.

Disposal of Radioactive Waste

Nuclear waste from commercial reactors, medical applications, defense industry, radioactive research poses a great threat to the scientific community as well as the public. The solutions to high-level nuclear waste are still debatable, both technically and ethically. There are numerous proposals for disposing high-level nuclear wastes. However, the most favored solution for the discarding of radioactive wastes is segregating out radioactive waste from man and biosphere for a long period of time such that any prospective release of radionuclides from the waste repository will not occur thereby resulting in unwarranted radiation exposure. The primary idea behind this solution is to use stable geological environments that can held the radioactive material for millions of years to provide an appropriate isolation capacity for the duration required.

The main reason for relying on such geological environments is based on the following chief consideration: 'Geological media is an entirely passive disposal system with no requirement for continuing human involvement for its safety. It can be abandoned after closure with no need for continuous surveillance or monitoring. The safety of the system is based on multiple barriers, both engineered and natural. The main one being the geological barrier.

Numerous alternatives for the disposal of long-lasting wastes have been examined over the years. Studies have been performed on concepts such as disposal in oceanic sub-seabed sediments and ultra-deep boreholes, besides on more exotic proposals such as disposal into geologic subduction zones, in polar icecaps and launching into space. All of these concepts have been found inadequate in terms of costs or risks, or unfeasible owing to political or legal restrictions. The following options have been aired sometime or the other. Each one of the options requires serious studies and technical assessments:

- Deep geological repositories.
- Under water disposal i.e. Ocean dumping mainly:
    - Seabed burial.
    - Sub-seabed disposal.
- Subductive waste disposal method.
- Transforming radioactive waste to non-radioactive stable waste.
- Dispatching to the Sun.

Most important problems due to legal, social, political and financial reasons have arisen in implementation due to:

- Environmental perceptions.

- Lack of proper awareness and education.
- 'Not-in-my-backyard' syndrome.
- 'Not-in-the-ocean' syndrome.
- Lack of proven technology.

Geologic Disposal

For handling long-lived radioactive waste, disposal in deep geological formations: under continental crust or under seabed has been recognized since 1957. Deep geological disposal involves introducing waste packages in underground structures dug in an impermeable geological medium having favourable properties in terms of its geological stability, hydrogeology, geochemistry and response to mechanical and thermal stress. The chosen medium must avoid areas of exceptional interest in terms of exploitable underground resources. The structures must be located at least 200 m below the ground surface to avoid the effects of erosion and human intrusion. Depositing the waste in deep geological sites will favor isolation of waste and keep them stable over thousands of years. Long-lived radioactive wastes are buried in deep sites unlike low-level radioactive waste which is normally disposed in near-surface facilities or abandoned mines.

Mostly, crystalline (granitic, gneiss) or argillaceous (clays) or salty or tuff are preferred for disposal of high-level radioactive waste. Provisional storage facilities, which permit cooling of the wastes over a few decades also, do exist in most of the countries. The majority of countries now believe deep geological disposal as the standard solution for final management of high-level and intermediate-level long lived waste. Granite; sedimentary formations, more especially clay beds; and salt are three types of geological formations on which most of the in progress research and studies are mainly focusing.

Underwater Disposal

Ocean-dumping

Generally, two associated methods of underwater disposal of Spent Nuclear Fuel (SNF) exist: dumping containers of radioactive waste into the ocean, and sub-seabed disposal. The main reason of underwater disposal of SNF is the same as any other type of SNF disposal, which is to isolate radioactive waste from human contact and the environment long enough for any release of radiation to become harmless.

USA, France, Great Britain, and many more industrialized countries have opted dumping into the oceans as the least expensive method for disposal of the radioactive wastes. Before the U.S. government banned or stopped sea dumping of spent nuclear fuel (SNF) around the 1980s, the United States reportedly dumped about 112,000 containers of nuclear waste at thirty locations in the Atlantic and Pacific oceans.

Although this practice has been prohibited by nearly all of the countries with nuclear programmes, the problem still persists. Russia, which presently controls sixty per cent of the world's nuclear reactors, continues to dispose of its nuclear wastes into the oceans. According to Russia's Minister of Ecology, it will continue to dump its wastes into the oceans because it has no other alternative method. It will continue to do so until it receives enough international aid to create proper storage facilities. In response, the United States has pledged money to help Russia, but the problem continues. Although radioactive waste has known negative effects on humans and other animals, no substantial scientific proof of bad effects on the ocean and marine life has been found. Therefore, some nations have requested that ocean-dumping should be carried on. Others argue that the practice should be banned until further proof of no harm is available.

Sub-seabed Disposal

Seabed disposal involves disposal of waste in deep oceans floor situated 5 km below the sea surface. These large tectonic plates are desert-like and are covered with hundreds of meters of thick sedimentary soft clay. Sub-seabed disposal is complexed and involves two associated methods: creating a repository with a possibility for retrieval of SNF, and everlasting burial.

Sweden and the United Kingdom consider creating a repository method, which permits for retrieval of SNF and might, contains admittance to the repository from land. A benefit of such a sub-seabed repository would be an increased capability to supervise SNF, in contrast to the dumping method. Another advantage could be a potential for access from land, which could let the repository method to avoid violating international bans against oceanic dumping, but accessing certain underwater locations by land might be impossible.

In Seabed Burial method, the mud flats at the bottom of the ocean is drilled to depths of the order of hundreds of meters. Inside these bore holes the radioactive material that are enclosed in containers are stacked vertically one over the other. Inconsistency of the rock and high local permeability limits the use of rocks in ocean as basement. Moreover, oceanic water has a mixing time of the order of a few thousand years which does not serve as a good barrier for long-lived radionuclides.

Experimental works have already established that clays have the property of holding on to several radioactive elements, including plutonium; hence, seepage of these elements into saline water is negligible. Rates of migration of these elements over hundreds of thousands of years would be of the order of a few meters. Hence, during such long times, radioactivity will lessen to levels below the natural radioactivity in sea water as a result of natural radioactive decay. The clays also contain plastic-like behaviour to form natural sealing agents. Lastly, the mud-flats have rather low permeability to water; hence, leaching probability is rather low. It may be noted that the method depends on standard deep-sea drilling techniques routinely practiced and sealing of the bore-holes. Core samples from about half a dozen vastly separated sites in the Pacific and Atlantic oceans have 'showed a continuous history of geological stillness over the past 50–100 million years'. However, there are several questions that still remain to be answered:

- Whether migration of radioactive elements through the ocean floor is at the same rate as that measured in the laboratories?
- What is the effect of nuclear heat on the deep oceanic-clays?
- What is the import on the deep oceanic fauna and waters above?
- In case the waste reaches the seabed-surface, will the soluble species (for example, Cs, Tc, etc.) be diluted to natural background levels? If so, at what rate?
- What occurs to insoluble species like plutonium?

- What is the possibility of radioactivity reaching all the way to the sea surface?
- In problems of accidents in the process of seabed burial leading to, say, sinking ships, to loss of canisters, etc. how does one recover the waste-load under such scenarios?
- What is the likelihood that the waste is hijacked from its buried location?
- Other problems include:
    - International agreement to consider seabed-burial as distinct from 'ocean-dumping'.
    - This method would be expensive to implement, but its cost would be an impediment to any future plutonium-mining endeavor.

Underwater disposal of SNF, specifically sub seabed disposal, can occur far from the coast of any state or nation and may perhaps thereby avoid the NIMBY ("not in my backyard") syndrome, but this result is not guaranteed considering existing laws and a popular belief that Earth's oceans are a global common.

Though the world tendency is in the direction of the option of land-based disposal, it is unsure whether confining repositories to land-based sites really assist in prevention of sea pollution. If radionuclides from a land-based repository leached out to the surface, they would be rapidly transported to the sea by surface water. The most essential thing is to isolate radionuclides from the biosphere as consistently as possible. If sub-seabed disposal results in more dependable isolation, sub-seabed disposal is the better safeguard against sea pollution. This method takes into consideration technological feasibility, protection of marine environments, and availability of international understanding. The United Nation's Convention on the Law of the Sea defines that a coastal state is granted sovereign rights to utilize all resources in water and under the seabed within its exclusive economic zone (EEZ), which can extend from the coast line up to 200 nautical miles (about 370 km) offshore.

A repository is proposed to be constructed in bedrock 2 km beneath the seabed. To utilize sub-seabed disposal within the EEZ, it is also proposed that waste packages would be transported through a submarine tunnel connecting land with the sub-seabed repository. Sea pollution by an accident during disposal work would be unlikely, because waste would never go through sea water during the work. The proposed method is a deviation of geologic disposal. Long-term monitoring is also possible by maintaining the access tunnel for some time after constructing artificial barriers. Everlasting burial is an additional method of sub-seabed disposal and might include at least three options: (1) Shallow penetration of the seabed, (2) deep burial through drilling, and (3) subduction involving tectonic dissipation of SNF. These options are not strictly separated categories. They are linked approaches that share alike features and could result in similar outcomes.

Subductive Waste Disposal Method

It is the latest disposal technology for nuclear wastes. This method is considered important because the material deposited will not return to the biosphere. It eliminates the material completely from the biosphere faster than it can return. Subductive method is safest, most sensible, most economical, most stable (for long term) and environmentally benign. It is considered safe because during subduction, the waste will be carried to the interior of the earth, diluted and dispersed. Subduction is a process whereby one tectonic plate slides beneath another and is ultimately reabsorbed into the mantle. The rate at which the plates subduct is 2.1 cm per annum which is faster than diffusion of radionuclides (1 mm annually) through the turbidite sediments. The movement of these plates is very slow, during which vertical fractures over a repository at the interface are sealed.

Transmutation of High-level Radioactive Waste

This means of high-level radioactive waste foresees that one may use transmutational devices, consisting of a hybrid of a subcritical nuclear reactor and an accelerator of charged particles to 'destroy' radioactivity by neutrons. 'Destroy' may not be the proper word; what is effected is that the fission fragments can be transmuted by neutron capture and beta decay, to produce stable nuclides. Transmutation of actinides involves several competing processes, namely neutron-induced fission, neutron capture and radioactive decay. The large numbers of neutrons produced in the spallation reaction by the accelerator are used for 'destroying' the radioactive material kept in the subcritical reactor. The scheme has not so far been demonstrated to be practical and cost effective.

Solar Option

It is proposed that 'surplus weapons' plutonium and other highly concentrated waste might be placed in the Earth orbit and then accelerated so that waste would drop into the Sun. Although theoretically possible, it involves vast technical development and extremely high cost compared to other means of waste disposal. Robust containment would be required to ensure that no waste would be released in the event of failure of the 'space transport system'.

There has been by a large agreement amongst the waste management community that deep geologic disposal represents the only possible route for ensuring sufficient long-term isolation of long-lived waste from the human environment and thus "closing the nuclear fuel cycle". Current debates within waste management programmes focus mostly on when and where disposal should take place, bearing in mind the need to fulfill ethical obligations, reduce present and future risks, ensure that other management options are given due consideration, and enhance societal participation in and ultimately acceptance of the disposal strategy.

Each nuclear facility in India has its own Near Surface Disposal Facility (NSDF). There are seven NSDFs currently operational within the country. These NSDFs in India have to address widely varied geological and climatologically conditions. The performance of these NSDFs is continuously evaluated to enhance the understanding of migration, if any and to adopt measures for upgrading the predictability over a long period of time.

Municipal Solid Waste

Municipal Solid Waste (MSW) can be defined as solid waste which includes all domestic refuse and non-hazardous wastes such as commercial and institutional wastes, street sweepings and construction debris. In some countries the solid wastes management system also handles human wastes such as night-soil, ashes from incinerators, septic tank sludge and sludge from sewage treatment plants. If these wastes manifest hazardous characteristics they should be treated as hazardous wastes.

Waste management practices differ for developed and developing nations, for urban and rural areas, and for residential and industrial producers. MSW is thus seen as primarily coming from households but also includes wastes from offices, hotels, shopping complexes/shops, schools, institutions, and from municipal services such as street cleaning and maintenance of recreational areas. Residential and commercial types of solid waste include food wastes, paper, cardboard, plastics, textiles, glass, metals, and ashes, special wastes like bulky items, consumer electronics, batteries, oil, tires and household hazardous wastes. Institutions types of solid waste include paper, cardboard, plastics, wood, food wastes, glass, metals, special wastes, hazardous wastes. Municipal services types of solid waste include Street sweepings, landscape and tree trimmings, general wastes from parks, beaches, and other recreational areas. Therefore, the major types of MSW are food wastes, paper, plastic, rags, metal and glass, with some hazardous household wastes such as electric light bulbs, batteries, discarded medicines and automotive parts.

In recent years the volume of waste has been increasing at an alarming rate, posing a formidable challenge to governments. The complexities and enormity of the challenges become evident when considering other waste types to be managed and these include industrial solid waste, municipal wastewater, industrial wastewater, storm water and hazardous waste. Often, different government agencies are mandated to manage different waste sectors. This fragmented approach to waste management, coupled with a lack of clear definition and delineation of the different waste types, makes an assessment of current waste management practices in most countries difficult.

Waste Management Strategies used in Municipal Solid Waste Management

Waste management is the collection, transport, processing, recycling or disposal, and monitoring of waste materials. Operations strategy can be viewed as part of a planning process that coordinates operational goals with those of the larger organization. Since the goals of the larger organization change over time, the operations structure must be designed to anticipate future needs. The operations capabilities of a firm can be viewed as a portfolio best suited to adapt to the changing product and service needs of a firm's customers.

The costs for solid waste management are high especially for collection, transportation, treatment and disposal, which are largely borne by city councils. Methods of collection of waste are either door-to-door or using containers or communal bins. All medium and large cities have administrative structures for providing collection services but often, cities in developing countries use non-compaction trucks for daily collection, with a few cities using compaction trucks and hauling trucks. The most common municipal waste management practices include: recycling/recovery, composting, incineration and land filling/open dumping. The operations strategy is a very important tool in the solid waste management practices and processes.

MSW may contain the following materials, which are considered recyclables: ferrous and non-ferrous metals, construction debris, scrap tires, paper/cardboard, plastics, textiles (including cloth and leather), glass, wood/timber, animal bones/feathers, waste oil and grease, cinders/ashes. In the middle-to-low-income cities, there exists a long-standing practice of informal source separation and recycling of materials. This has led to the development of enterprises for the gathering, trading and reprocessing of materials. For example Mukuru Recycling project which started in 1991 to help men and women scavengers sell recyclable waste to industries. The national ministries support waste recovery and recycling activities at city level although many of these are family businesses. However, since industries would only be interested to use recycled materials when they cost less than the virgin materials, the practice of recycling is so market-driven that recycling has become selective. The disposal of those unselected recyclables remains a problem.

Informal waste separation or waste picking takes place in three ways: At source - this is in large urban areas, e.g., commercial areas or residential areas with apartments/high-rise buildings for high income earners. Here waste pickers sort out the waste before the authorized collection vehicle arrives. During collection, when the collectors segregate recyclable materials during loading and store them inside the truck or on the sides of the vehicles. At the disposal site – where the waste pickers often live on or near the dumps. However, they risk the danger of potential slides and fires. While waste picking means survival for waste pickers the methods of uncontrolled waste picking can reduce the efficiency of the formal collection system and can be detrimental to health due to exposure to biological pathogens.

Composting is not well practiced. Waste materials that are organic in nature, such as plant material, food scraps, and paper products, can be recycled using biological composting and digestion processes to decompose the organic matter. The resulting organic material is then recycled as mulch or compost for agricultural or landscaping purposes Household organic wastes, including wastes from the restaurants, are often collected for animal feed. But these are either not working or are not operating at full capacity for a number of reasons, such as: High operating and maintenance costs, poor maintenance and operation of facilities, Incomplete separation of non-compostables, such as, plastics and glass, high cost of compost compared to commercial fertilizers.

Another waste treatment method that is practiced is incineration where 90 percent of non-recyclable municipal solid waste is incinerated. Final disposal of waste is at landfills where 10 percent of non-recyclable municipal solid waste is deposited. Singapore has four government-owned and operated incinerators for the disposal of solid waste that is not recycled. However, controversy remains over the soundness of incineration as a waste treatment technology because of greenhouse gas emissions from incinerators. Incineration has been completely banned under the new law on solid waste management. The practice of informal incineration or open burning is, however, still prevalent, not only in the rural areas where waste collection is rare but also in peri-urban and urban areas.

The popular meaning of 'recycling' in most developed countries refers to the widespread collection and reuse of everyday waste materials such as empty beverage containers. These are collected and sorted into common types so that the raw materials from which the items are made can be reprocessed into new products. Material for recycling may be collected separately from general waste using dedicated bins and collection vehicles, or sorted directly from mixed waste streams.

Landfills are generally the cheapest and most common disposal method for municipal solid waste. Disposing of waste in a landfill involves burying the waste, and this remains a common practice in most countries. Landfills were often established in abandoned or unused quarries, mining voids or borrow pits. A properly designed and well-managed landfill can be a hygienic and relatively inexpensive method of disposing of waste materials. Older, poorly designed or poorly managed landfills can create a number of adverse environmental impacts such as wind-blown litter, attraction of vermin, and generation of liquid leachate. An exception is a large city like Singapore, which faces rising disposal costs due to exhaustion of traditional disposal sites, stricter environmental controls and greater waste quantities, thus requiring other methods like incineration to reduce the volume of waste for final disposal. In the other developing countries, open dumping is the common practice, i.e., municipal solid waste is dumped on swamplands and low-lying areas, which are eventually reclaimed for development. The problems associated with landfills, even with those that are clay-lined, include high water table, groundwater contamination and gas migration.

Incineration is a disposal method in which solid organic wastes are subjected to combustion so as to convert them into residue and gaseous products. This method is useful for disposal of residue of both solid waste management and solid residue from waste water management. This process reduces the volumes of solid waste to 20 to 30 percent of the original volume. Incineration and other high temperature waste treatment systems are sometimes described as "thermal treatment". Incinerators convert waste materials into heat, gas, steam and ash. Incineration is common in countries such as Japan where land is more scarce, as these facilities generally do not require as much area as landfills. Waste-to-energy (WtE) or energy-from-waste (EfW) are broad terms for facilities that burn waste in a furnace or boiler to generate heat, steam and/or electricity.

Formulation of the Municipal Solid Waste Management Operations Strategy

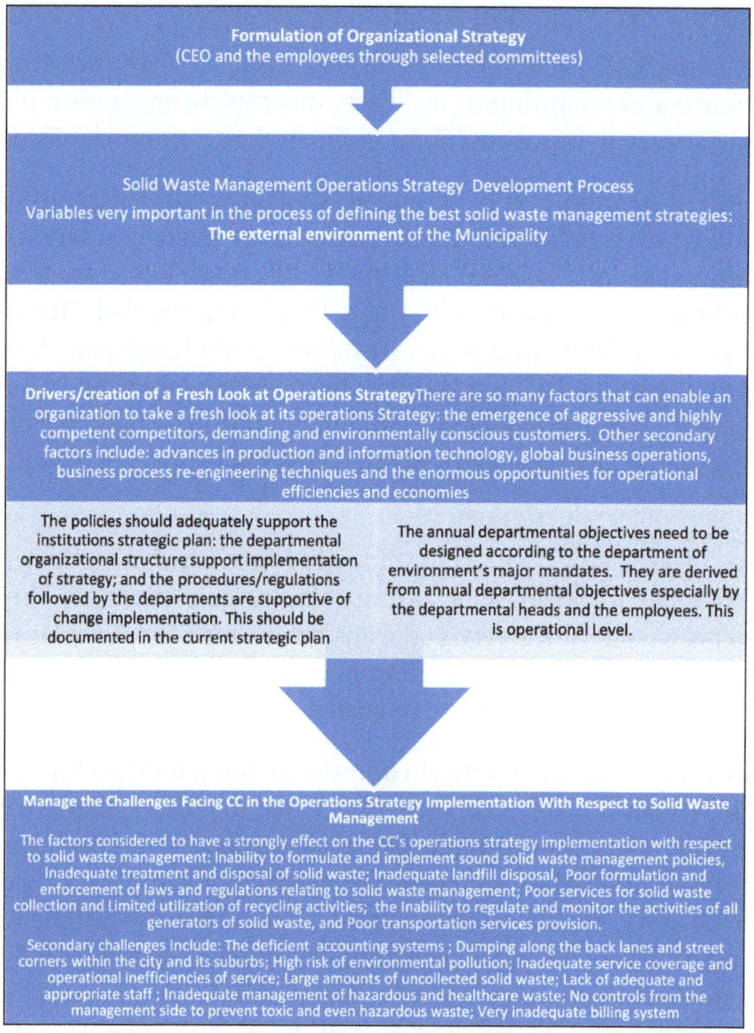

Formulation of strategies used in municipal solid waste management.

Operations strategy is the "HOW" in any corporate and market strategy. Operations strategy is no longer a tool for continuous improvement and sustainable competitive advantage in the manufacturing sector only, since it can be now applied in the service industry and public organizations.

The operation strategies used in solid waste management can be modeled using a process chart. From the model, the formulation of organizational strategy must be done by the CEO and the employees through selected committees. The formulation of organizational strategy should be followed by setting of Annual Objectives in Solid Waste Management. The annual objectives includes to: improve public health of the people; improve the environment; and maintain public cleanliness in order to keep public places aesthetically acceptable; by ensuring the proper storage, collection, transportation, safe treatment and disposal of solid waste. This driven by the annual departmental objectives designed according to the department of environment's major mandates. They are derived from annual departmental objectives especially by the departmental heads and the employees. This is operational Level. The policies adequately support the institutions strategic plan: the departmental organizational structure support implementation of strategy; and the procedures/regulations followed by the departments are supportive of change implementation. This should be documented in the current strategic plan.

There are so many factors that can enable an organization to take a fresh look at its operations Strategy. The different factors that impact on the operations strategy are: most managers felt that the emergence of aggressive and highly competent competitors, demanding and environmentally conscious customers. Other secondary factors include: advances in production and information technology, global business operations, business process re-engineering techniques and the enormous opportunities for operational efficiencies and economies. The external environment of the municipality's trends in the market; changes in customer wants and expectations (demanding speed of delivery, high quality, and low price); Identifying the company's strengths, special skills of workers, such as expertise in providing customized services or knowledge of information technology; the trends in the political environment changes in the political climate—local, national, and international) and forming partnerships with international firms.

Investment on scanning the environment of operation before developing the solid waste management operations strategy is important. As one invests on time and effort in implementing its operations strategy after the environmental scanning and formulation of operations strategy, there is need to invest in its implementation. There are a number of challenges which needs to be managed during the implementation of operations strategy with respect to solid waste management.

The factors considered to have a strongly effect on the municipal solid waste strategy implementation include: Inability to formulate and implement sound solid waste management policies, Inadequate treatment and disposal of solid waste; Inadequate landfill disposal, Poor formulation and enforcement of laws and regulations relating to solid waste management; Poor services for solid waste collection and Limited utilization of recycling activities; the Inability to regulate and monitor the activities of all generators of solid waste, and Poor transportation services provision. The secondary challenges include:

- The deficient accounting systems.
- Dumping along the back lanes and street corners within the city and its suburbs.

- High risk of environmental pollution.
- Inadequate service coverage and operational inefficiencies of service.
- Large amounts of uncollected solid waste.
- Lack of adequate and appropriate staff.
- Inadequate management of hazardous and healthcare waste.
- No controls from the management side to prevent toxic and even hazardous waste.
- Very inadequate billing systems.

E-Waste

Electronic waste or 'e-waste' in general refers to electrical and electronic equipment (EEE) that has ceased to be of any value to its owners. European Union WEEE Directive 2002/96 which, is one of the earliest legislation in the world to have recognised this specialised waste stream, defines EEE as equipment which is dependent on electric currents or electromagnetic fields in order to work properly and equipment for the generation, transfer and measurement of such currents and fields and designed for use with a voltage rating not exceeding 1000 volts for alternating current and 1500 volts for direct current. WEEE is defined as EEE which is waste ('waste' means any substance or object which the holder disposes of) including all components, subassemblies and consumables which are part of the product at the time of discarding.

E-waste encompasses a broad and growing range of electronic devices ranging from large household devices such as refrigerators, air conditioners, cell phones, personal stereos, and consumer electronics to computers, which have been discarded by their users. Although e-waste broadly refers to electrical and electronic equipment that have lost valuable use to the current user what encompasses e-waste that requires to be managed varies according to the regional and national legislations. For example, while EU WEEE recognizes 10 categories of e-waste that ranges from large household equipment to toys, sports and medical equipment. IT and telecommunication equipment and consumer electrical and electronics including fluorescent and mercury containing lamps. Despite the wide and varied definition on what constitutes e-waste, globally it has been acknowledged that this is one of the fastest growing streams of waste that needs immediate attention.

This e-waste contains a diverse range of materials. Most studies examine five categories of materials in e-waste viz. ferrous metals, non-ferrous metals, glass, plastics and others. Iron and steel account for almost half the total weight of WEEE, followed by plastics (21% of weight), and non-ferrous metals including precious metals (13%, of weight of which copper accounts for nearly 7%). It also contains around 2.7% of pollutants. Nearly 60% of materials like iron, steel, glass and plastic along with recoverable valuable metals like gold in e-waste are recyclable.

Sources of E-waste

Electrical and electronic equipment include a large gamut of products ranging from small household equipment, to toys, to large equipment like refrigerators, washing machines, IT equipment, mobiles, office equipment, medical equipment etc. It is these equipment that become e-waste at the end of their useful life. So the sources of e-waste are also very large ranging from individual users, to households, to organizations. Therefore, the generators of e-waste are classified broadly as individual (which also include households) and bulk consumers.

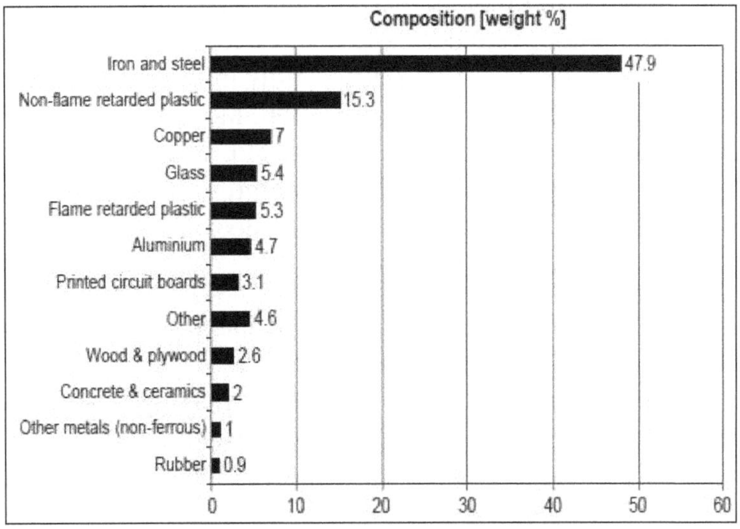

Typical material fractions in WEEE.

Classification of e-waste is important for compiling e-waste data regarding its generation and management. With regards to classification of e-waste it varies from country to country and so there is no generic system. The United Nations University (UNU) classification also known as UNU-keys that classifies e-waste based on products of similar function, comparable material composition (in terms of hazardous substances and valuable materials) average weight, end-of life characteristics and life span distribution is regarded as most comprehensive classification. It has 54 categories of e-waste that are grouped under ten primary categories.

The Challenge of E-waste Management

Presence of nearly 60% of recyclable material in e-waste has made the global, transboundary trade in e-waste from the developed countries to the developing countries a profitable business. Despite opposition from NGOs world over, this global trade in e-waste between the developed and developing countries has been increasing annually. The recyclability of e-waste together with the presence of potentially toxic pollutants poses a waste management challenge in the developing countries, where the resources in these wastes are extracted using archaic methods that cause damage to the workers and the environment. This e-waste that is potentially damaging if mismanaged is growing in quantity globally.

The number of appliances put onto the market every year is increasing both in post-industrialized and industrializing countries, thereby contributing to the increased generation of e-waste globally. E-waste contains both toxic and valuable materials. On the one hand the presence of toxic substances makes it environmentally challenging to handle it, while the presence of valuable substance encourages an international trade in e-waste. The growing quantity of e-waste, together with its toxic nature has made the management of e-waste a global challenge.

In the developed countries, the increasing quantity of e-waste is attributed to higher cost of repair of electronic equipment together with the decreasing cost of equipment, which makes it economically viable to replace rather than repair. Planned obsolescence practiced by EEE (electrical and electronic equipment) producers has further enhanced the consumer's desire to replace products. Planned obsolescence is a marketing strategy that creates long-term sales volume by decreasing the useful life of technology, thereby increasing the rate of purchase of new technology. This planned obsolescence of consumer products leads to more frequent purchases not out of personal choice but as a consequence of the products' purposefully limited durability, thus increasing the rate of consumption and ultimately the waste generation. Most often the e-waste thus generated in developed countries ends up in landfills or is incinerated. Puckett and Smith reported that the e-waste produced in the US ended in landfills along with other municipal solid waste. Similar behaviour was also observed in the European Union until a decade back.

With the decreased availability of landfill space and increased recognition of the toxic nature of e-waste, regulations have been promulgated to manage them. The promulgation of regulations has made it expensive to recycle e-waste in most developed countries and hence they have chosen the alternate cheaper route of managing this e-waste by exporting it to the developing countries of the South. It has been reported that the United States e-waste recycling industry once declared that around 80% of the e-waste they received was exported into Asia, and around 90% of it went to China.

In the developing countries the increasing quantity of e-waste is not only due to the economic growth and related change in consumer behaviour, but also due to the import of these wastes from developed countries for recycling. Developing countries tend to view e-waste as a resource for income generation, as opposed to 'waste' owing to the availability of large-scale unskilled labour in the informal sector that is involved in the dismantling and recovery process of valuable metals present in them. This large informal sector, engaged in resource recovery from e-waste, does so by adopting environmentally unsound practices. The nature of e-waste, along with the lack of proper regulation to handle it in developing countries, together with it being looked upon as a resource for income generation, makes e-waste management a challenging issue in these countries.

Management of E-waste

Understanding the nature of waste plays an important role in the management of waste. With regards to e-waste the source of e-waste is in itself very vast compared to the other wastes whose sources are more specific (e.g. Municipal solid waste, hospital waste etc.). Also the presence of both valuable recyclable material and toxic material together makes it essential to have a combination of management techniques that is unlike other wastes.

World over the management of e-waste is based on the principle of extended producer responsibility (EPR). "EPR is defined as an environmental protection strategy that makes the manufacturer of the product responsible for the entire life cycle of the product and especially for the take back, recycling and final disposal of the product". It was Lindhqvist, who first proposed this environmental policy strategy to manage e-waste, covering five parameters to be considered when designing an EPR based e-waste management system, viz. legal regulation, system coverage, system financing, producer responsibility, compliance. He studied e-waste management in different countries through this framework and showed how these parameters were closely interlinked and play an important role in the effective management of WEEE.

By shifting the responsibility of financial and infrastructure burdens to tackle the waste from the municipality to the producers, EPR internalizes environmental externalities to a large extent. This linking of the manufacture phase of the product with its disposal by EPR encourages the manufacturers to go for better product design to enable easy upgrading and recycling according to Tojo.

The Swiss ORDEE (The Return, the Taking Back and the Disposal of Electrical and Electronic Equipment) 1998 has been the pioneering legislation for e-waste management. The European Union's Waste Electrical and Electronic Equipment Directive which promotes collection and recycling of e-waste and the Restriction on Use of Certain Hazardous Substances in electrical and electronic equipment RoHS - Directive are two important legislation pertaining to e-waste management in the EU countries. Many countries in Asia like Japan, South Korea, and Taiwan also have their own national regulation to tackle the challenge of managing e-waste. The adoption of EPR based approach to manage e-waste varies geographically and ranges from fully voluntary (Switzerland) to mandatory (EU) systems. Also, this policy approach in the developed countries has used market instruments like advanced recycling fees (ARF), deposit refund schemes, etc., to not only secure financial aid to operate such schemes, but also to enroll the participation of consumers. Despite such measures the dumping of e-waste from developed countries to developing countries continues.

Lack of regulation in developing countries that handle large volumes of e-waste imports is a serious challenge to its management. Many developing countries like China and India have now implemented e-waste legislation in line with the EU WEEE directive. Researchers like have recommended the need to mandate financial and collection responsibilities apart from legal responsibilities on producers in developing countries EPR policy for e-waste management to ensure proper e-waste treatment.

In developing countries e-waste is still potentially regarded to have value and is not disposed of for free. Therefore, market based instruments like ARF, tax credits, and deposit refund schemes that could be leveraged for participation of stakeholders in e-waste management in developing countries have been proposed by researchers while analysing the context in these countries. Also in the developing countries there is a large group of participants in the informal sector that make a living by resource extraction from this waste through primitive backyard recycling who cannot be ignored in an e-waste management system. The integration of the informal sector with formal e-waste recyclers by involving them only in collection, for effective e-waste recycling in developing countries where informal recycling of e-waste is high has been recommended by researchers.

E-waste Treatment

Since e-waste is a mixture of valuable material that is recoverable and recyclable with toxic substance that needs to be safely disposed its treatment is also complex. E-waste requires both labour-intensive manual segregation along with capital intensive technical processes for the separation of toxic waste. Handling e-waste by beginning with manual dismantling has been recommended as the best starting process for its treatment. This prevents mixing up the different constituent materials which otherwise reduces its value and increases the challenges in recovering material. The dismantled e-waste is separated into glass, copper, steel, aluminium, plastic, printed circuit boards, etc. The total content of printed circuit board in e-waste is 3-5% by weight, while metals, plastic, and glass, constitute the remaining 95-97%. The hazardous components like capacitors and batteries, CRT screens, CFC gases, light bulbs and batteries are also separated and removed at this stage.

Once the critical toxic compounds are removed the e-waste is subjected to mechanical process. This mechanical processing which is normally a large scale operation enables increase of recyclable materials in a dedicated fraction and also helps to further separate hazardous materials. Typical components of a mechanical processing plant are crushing units, shredders, magnetic- and eddy-current- and air-separators. The gas emissions are filtered and effluents are treated to minimize environmental impact. High-pressure compaction and cement solidification can also be used for the treatment of printed wire board (PWBs) into safe forms for co-disposing with MSW.

The final step in e-waste recycling is refining. This is an energy and capital-intensive process. Most of the fractions obtained here are refined to be sold as secondary raw materials. During the refining process, attention is paid to metals, glass and plastic. At the end of refining and after extraction of valuable fractions the residue, which is usually non-usable and toxic, is disposed in specially designed disposal facilities for hazardous waste.

Biomedical Waste

Biomedical waste, also known as infectious waste or medical waste is defined as waste generated during the diagnosis, testing, treatment, research or production of biological products for humans or animals. Biomedical waste includes syringes, live vaccines, laboratory samples, body parts, bodily fluids and waste, sharp needles, cultures and lancets.

Improper management of waste generated in health care facilities causes a direct health impact on the community, the health care workers and on the environment. The waste generated in these institutions essentially consists of solids and liquid, which may be hazardous, infectious and non-infectious. It has been estimated that up to 85% to 90% of the waste generated in hospitals is non-infectious (free with any body fluids, which is similar to domestic waste). It is the remaining 10% to 20% of waste that is of concern because it is hazardous and infectious. In addition, waste that is un-segregated and not treated in the right manner would cause environmental pollution affecting the health of the community. From waste audits done at several hospitals by a few NGOs, arrived at some figures, which can now be used and extrapolated for the whole country. These audits must be conducted only after adequate training on waste segregation is given to health care institutions.

Proper handling, treatment and disposal of biomedical wastes are important elements of health care office infection control programme. Correct procedure will help protect health care workers, patients and the local community. If properly designed and applied, waste management can be a relatively effective and an efficient compliance-related practice. Proper collection and segregation of biomedical waste are important. At the same time, the quantity of waste generated is equally important. A lesser amount of biomedical waste means a lesser burden on waste disposal work, cost saving and a more efficient waste disposal system. Hence, health care providers should always try to reduce the waste generation in day-to-day work in the clinic or at the hospital.

Sources of Biomedical Waste

The source of biomedical waste is the place or the location at which biomedical waste has been generated. The source of biomedical waste is classified into two types based on the quantity of waste generated. They include major and minor source. Major source generates more amount of biomedical waste compared to minor source and also there is regular generation of biomedical waste in the major source which includes government hospitals, private hospitals, nursing home and dispensaries. Minor source includes physicians and dental clinics. Figure shows the details of the various source of biomedical waste generation.

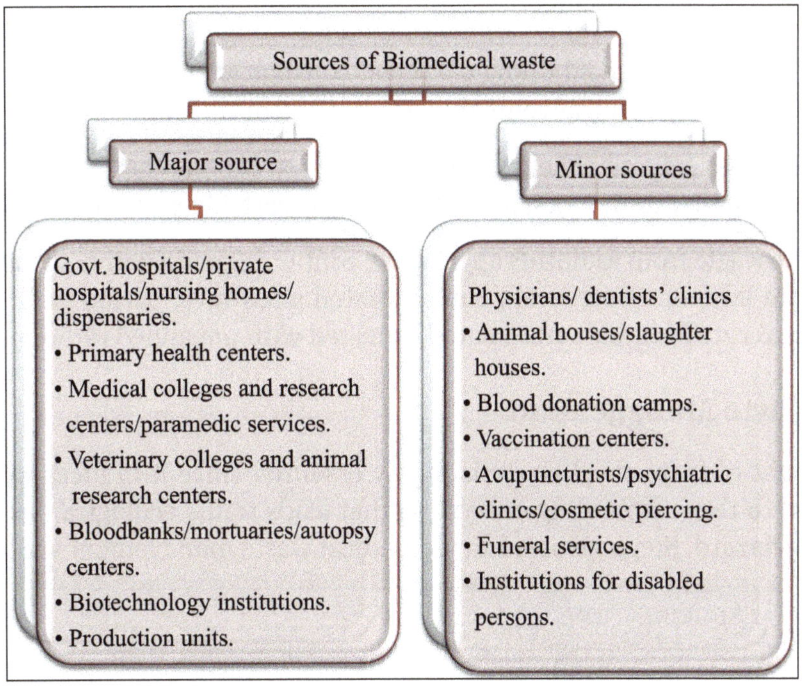

Classification of Biomedical Waste

The classification of the biomedical waste is carried out based on its characteristics, source of generation and the level of hazard to the environment. The biomedical waste is classified into two types: Non-hazardous waste and Hazardous waste.

Non-hazardous Waste

About 75% to 90% of biomedical waste characteristics were similar to that of domestic waste and are non-risky in nature. This waste is generated mainly from the organization and maintenance of hospital and health care centers.

Hazardous Waste

The remaining 10 – 25% of biomedical waste falls under the hazardous waste categories. The hazardous waste contains infectious characteristics of about 15% - 18 % and toxicity characteristics of about 5% - 7%. The various hazardous wastes includes:

- Infectious waste: Waste containing pathogens; e.g. excreta; laboratory cultures; isolation wards waste; swabs, materials, or equipments that have been in contact with infected patients.
- Pathological waste: Human tissues or fluids e.g. body parts; blood and other body fluids; fetuses.
- Pharmaceutical waste: Waste containing pharmaceuticals; e.g. pharmaceuticals that are expired or no longer needed; contaminated pharmaceuticals (bottles, boxes).
- Genotoxic waste: Waste containing cytostatic drugs (often used in cancer therapy)/ genotoxic chemicals.
- Chemical waste: Waste containing chemical substances e.g. laboratory reagents; film developer; disinfectants and solvents that are expired or no longer needed.
- Wastes with high content of heavy metals: Batteries, Broken thermometers, blood pressure gauges, Pressurized containers, gas cylinders, gas cartridges, aerosol cans.
- Radioactive waste from radiotherapy: Waste containing radioactive substances e.g. unused liquids from laboratory research; contaminated glassware, packages or absorbent paper; urine and excreta from patients treated or tested with uncapped radionuclide.

Biomedical Waste Management

Proper management of biomedical waste is highly essential since it induces various risks to the human health and to the surrounding ecosystem that leads to the ecological hazard, professional hazard and public hazard. Steps involved in biomedical waste management are shown in figure.

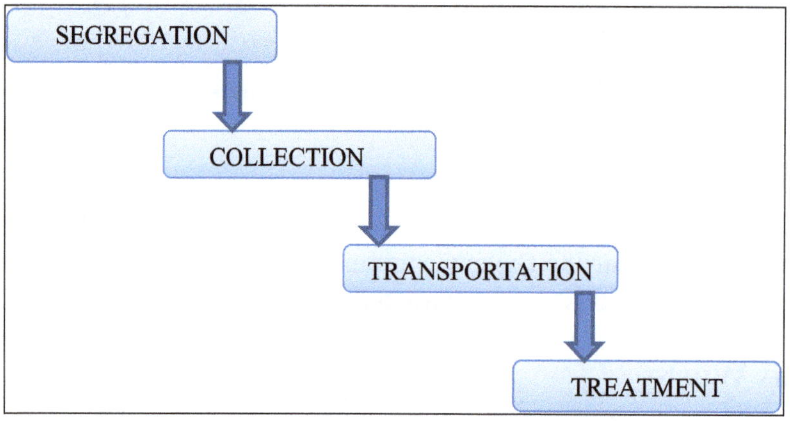

Steps involved in biomedical waste management.

Segregation

To avoid mixing of the biomedical waste with other, a container should be set to the side with colour coding bags at the point of generation. The sorting or separation of waste into different categories is referred as segregation. Segregation will decrease or minimize the risks in addition to rate of managing and disposal. Segregation is the most important and critical step in bio-medical waste management. Only, effective segregation can confirm the effective bio-medical waste management.

How does Segregation Help?

Segregation plays an effective role in handling and treatment of waste. It reduces the quantity of waste and if done effectively, it can avoid the mixing of biomedical waste with any other type of waste especially municipal waste. Segregation will avoid the reuse of certain biomedical waste like used syringes, needles and other plastics. Some materials like plastics can be recycled after proper disinfection and these can be reused for non-food grade products.

During segregation process, the biomedical waste must be separated under the following categories shown in Table. Category no. 1 includes the Human anatomical waste in which the human tissues, organs, body parts are considered. Animal waste falls under the Category No. 2. It includes Animal tissues, organs, body parts, carcasses, bleeding parts, fluid, blood and experimental animals used in research, waste generated by veterinary hospitals and colleges, discharges from hospitals, animal houses. Category No. 3 is the Microbiology and Biotechnology waste which contain Wastes from laboratory cultures, stocks or specimen of live microorganisms or attenuated vaccines; human and animal cell cultures used in research; infectious agents from research and industrial laboratories; wastes from production of biologicals, toxins and devices used for transfer of cultures. The Category No. 4 includes waste Sharps in which Needles, syringes, scalpels, blades, glass, etc. that may cause puncture and cuts. This includes both used and unused sharps. Discarded Medicine and Cytotoxic drugs falls under the Category No 5 which consists of wastes comprising of outdated, contaminated and discarded medicines. The soiled waste is included in the Category No. 6 containing items contaminated with body fluids including cotton, dressings, soiled plaster casts, lines, bedding and other materials contaminated with blood.

Table: Categories of Waste.

Waste Category	Type of Waste
Category No. 1	Human Anatomical Waste
Category No. 2	Animal Waste
Category No. 3	Microbiology and Biotechnology Waste
Category No. 4	Waste Sharps
Category No. 5	Discarded Medicine And Cytotoxic Drugs
Category No. 6	Soiled Waste
Category No. 7	Solid Waste
Category No. 8	Liquid Waste

Category No. 9	Incineration Ash
Category No.10	Chemical Waste

Category No. 7 is the solid waste which includes waste generated from disposable items other than the waste sharps such as tubing, catheters, intravenous sets, etc. Liquid waste falls under the category no. 8, it consists of waste generated from the laboratory and washing, cleaning, housekeeping and disinfecting activities. Category No. 9 includes incineration ash i.e., ash from incineration of any biomedical waste. Chemical Waste falls under Category No. 10 and consists of Chemicals used in production of biologicals, chemicals used in disinfection and as insecticides etc.

Collection and Storage

The collection of biomedical waste involves the installation of different colour coded containers for biomedical wastes obtained from varying sources. The containers/bins should be placed in a location so that 100% collection is achieved. The bins and bags that hold the biohazard symbol as shown in Figure represents the nature of waste. The symbols in biomedical waste management is generally used as a warning to take precautions while exposing to those substances. The biohazard symbol was developed by the Dow Chemical Company in 1966 for their containment products.

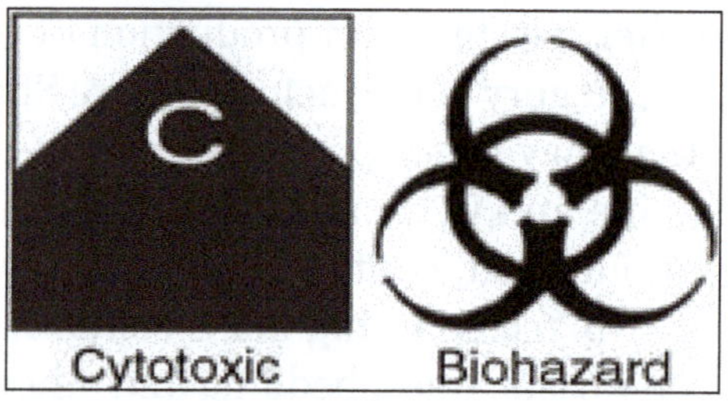

Symbols

Subsequent to collection, the biomedical waste is stored in specific containers and stored in a proper place. The extent of storage should not exceed beyond 8-10 h in big hospitals containing more than 250 bedded and 24 h in nursing homes. Each container must be clearly labelled with the location being mentioned in them. The purpose of labelling is to trace the waste at the source. Storage spot must be clear with a warning sign.

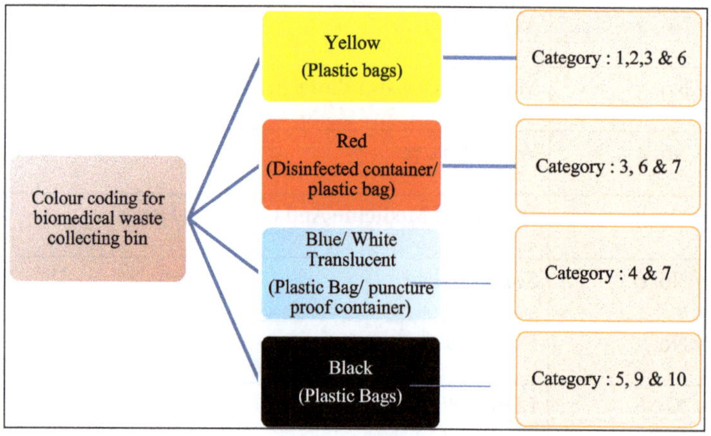

Collection of biomedical waste in a colour coded boxes.

Collection of the biomedical waste was carried out in its specific coloured bags. In the yellow colour bags, the categories 1,2,3 and 6 waste will be collected and this bags are made up of plastic materials. The Red bags are made up of disinfected container or plastic in which Category 3, 6 and 7 waste will be collected. The Blue/White Translucent bags collect Category 4 and 7 waste which is made up of Plastic/ puncture proof container. The black coloured plastic bags are used for the collection of waste under category 5, 9 and 10. Previous figure shows the collection of biomedical waste in the colour coded boxes.

Transportation

The collected wastes are transported in trolleys or in enclosed wheelbarrow for treatment. The operator should ensure to avoid manual loading. The bags/Container containing biomedical wastes must be tied/lidded before hauling for treatment. Vehicles used for transporting should be special to avoid contact to, and direct contact with the operator, scavengers and the public. While transporting the containers, it must be properly enclosed. The effects of traffic accidents should be incorporated in the design, and the driver must be trained in the actions which must be followed in case of an accidental spillage. The interior of the containers should also be rinsed thoroughly.

- Trolleys: The use of trolleys will make the elimination of infectious waste possible at the source itself, instead of accumulation a new category of waste.
- Wheelbarrows: Wheelbarrows are used to transfer the waste from the point source to the collection centres. There are two types of wheelbarrow – covered and open. Wheelbarrows are made of steel and provided with two wheels and a handle. Open dumping should not be done. Only packed waste (in plastic bags) should be carried. To prevent corrosion, care should be taken to prevent the liquid waste from spilling into the wheelbarrow. Wheelbarrows also come in various sizes depending on the utility.
- Chutes: Chutes are vertical conduits provided for easy transportation of biomedical waste vertically in case of establishment with more than two floors. Chutes should be produced from stainless steel. It should have a self-closing lid. These chutes have to be sterilized on a daily basis with formaldehyde vapours. The linen that is contaminated with blood or other body fluids from each floor must be bundled in soiled linen or in plastic bags before expelling into the chute.

  Alternately, elevators with mechanical winches or electrical winches can be used to bring down waste containers from each floor. Chutes are essential to keep away from horizontal transport of waste thereby diminishing the routing of the waste within the premises and hence reducing the risk of secondary contamination.
- Dustbins: It is very important to calculate the amount of waste generated at each point. Dustbins should be of such capacity so that it can be placed at this specific site and that they do not overflow between each cycle of waste collection. Dustbins have to be cleaned subsequently at each cycle of clearance of waste with disinfectants. Dustbins can be wrinkled with plastic bags, which are chlorine-free, and colour coded as per the law.

Treatment and Disposal

Before its final disposal of biomedical waste, it must be disinfected. Anatomical waste can be disposed by deep burial. Syringes to be cut (with hub cutters) and chemically disinfected with 1% bleaching powder solution at source of generation before final disposal into sharps pit. Infected plastics to be chemically disinfected or autoclaved, shredded and recycled and sent for final disposal into municipal dumps.

Incineration

Most of the hazardous biomedical wastes were treated by the method of incineration to reduce organic and combustible waste to inorganic incombustible matter. Incineration is a high temperature, dry oxidation process that results in significant reduction of waste volume and weight. Wastes that cannot be reused recycled or pose problem in disposing in landfills are treated by incineration. Examples of wastes that cannot be incinerated are chemical wastes, wastes containing high mercury or cadmium (broken thermometers, second-hand batteries, and lead lined wooden panels, sealed ampules or ampules containing heavy metals), silver salts, pressurized gas containers, photographic or radiographic wastes, halogenated plastics such as PVC.

The advantages of incinerator include high reduction of waste volume in addition to good disinfection competence. It helps to save the space in the landfill. The ash generated can be disposed of safely in the landfills. The major disadvantage of incineration includes high operating cost as they are energy intensive process. Also it releases a huge amount of atmospheric pollutants. The need for cyclic removal of slag and dirt, inadequacy in demolishing anti-thermal chemicals and drugs such as cyto toxic are its other disadvantages.

Autoclaving of Biomedical Waste

Autoclave treats the bio-medical waste through the mechanism of disinfection. The biomedical waste was subjected to following temperature and pressure based on its residence time:

- If the autoclave residence time is not less than 60 minutes, the temperature should not be less than 121 °C with the pressure of 15 pounds per square inch (psi).
- If the autoclave residence time is not less than 45 minutes, the temperature should not be less than 135 °C with the pressure of 31 pounds per square inch (psi).
- If the autoclave residence time is not less than 30 minutes, the temperature should not be less than 149 °C with the pressure of 52 pounds per square inch (psi).

While operating a gravity flow autoclave, biomedical waste is subjected to all three condition, whereas in vacuum autoclave, the biomedical waste is first subjected to one pre-vacuum autoclave (minimum) to purge the autoclave of all air. Succeeding this first and second conditions are applied. Bacillus stearothermophilus spore dials or spore strips with at least 1 × 104 spores per ml.is used as biological indicator of autoclave. The operating conditions of autoclave include a residence time less than of 30 minutes, temperature less than 121 °C or a pressure must be less than 15 psi. On reaching certain temperature, the chemical indicator strip/tape changes colour that indicates the attainment of specific temperature. It may be essential to use more than one strip at various locations on the waste package to ensure the effectively autoclaving of inner content of the waste in the package.

Biomedical Liquid Waste

Before disposing the liquid form of biomedical waste into the sewer, it must be treated. Pathological waste after being treated with chemical disinfectants are flushed into the sewage system. Likewise, the chemical waste is neutralized with suitable reagents and then either flushed or treated in the sewage treatment plant. Mostly they are neutralized and dumped in sewer network. Highly skilled operators are required for this technique as it involves handling of hazardous substances. The biomedical waste effluent generated from the various source should conform to the following limits shown in Table. Environment (Protection) Act, 1986 prescribes the discharge limits of these waste into public sewers.

Table: Disposal standard for biomedical waste.

Parameters Permissible limits	
pH	6.5-9.0
Suspended solids	100 mg/L
Oil and grease	10 mg/L
Biochemical Oxygen Demand (BOD)	30 mg/L
Chemical Oxygen Demand (COD)	250 mg/L

Microwave Treatment

Microwave treatment uses a frequency and wavelength of 2450 MHz and 12.24 cm, respectively for the destruction of microorganisms. The infectious contaminants in water with biomedical waste are destroyed by heat conduction when it is rapidly heated by the microwaves. By bacteriological and biological tests, the efficiency of the microwave disinfection was ensured regularly. The biomedical waste is evenly heated to a temperature of 97-100 °C by means of microwaves in treatment chamber. Treatment of biomedical waste by microwaving can be carried out in the source itself. No shredding is required for microwave treatment of waste.

Most infectious wastes except body parts, human organs, infected animals carcasses and metal objects are suitable for treatment by microwave technique. This method shows good disinfection competence with good waste shrinking capacity. Similar to incineration this method also involves high operating costs. It is an eco-friendly process with potential operation and maintenance problems.

Deep Burial

Deep burial process is done in pits or trench of about 2 meters deep. The pits are half filled with waste, 50 cm soil and then with waste. The pits are covered with galvanized iron/wire meshes. When wastes are added to the pit, a layer of 10 cm of soil shall be added to enclose the waste. The deep burial site should be impermeable with no shallow well in the nearby area. The pits should be away from the habitation to avoid infection to surface or ground water. The site selected should not be a flooding or eroding zone and should be approved by the authority.

Inertization

Assimilation of waste with cement and other substances before disposal is called inertization process. This decreases the risk of entry of toxic substances into the surface or groundwater. A typical percentage of the mixture is 65% pharmaceutical waste, 15% cement and 5% water. A homogenous mass is created and cubes or pellets are produced and then stored. This process is economical and not suitable for infectious waste.

Table: Treatment and disposal of biomedical waste.

Waste Category	Treatment and Disposal Option
Category No. 1	Incineration/deep burial.
Category No. 2	Incineration/deep burial.
Category No. 3	Local autoclaving/microwaving/incineration.
Category No. 4	Disinfecting (chemical treatment/autoclaving/microwaving and mutilation/shredding).
Category No. 5	Incineration/destruction and drugs disposal in secured landfills.
Category No. 6	Incineration/autoclaving/microwaving.
Category No. 7	Disinfecting by chemical treatment/autoclaving/microwaving and mutilation/shredding).
Category No. 8	Disinfecting by chemical treatment and discarding it into drains.
Category No. 9	Disposal in municipal landfill.
Category No.10	Chemical treatment and release into drains for liquids and protected landfill for solids.

Table above shows the treatment and disposal method of the different categories of biomedical waste. The process such as incineration, deep burial, disinfecting process and municipal landfill disposal will be carried out. Category 1, 2, 3, 5 & 6 can be incinerated. Disinfecting process includes chemical treatment, autoclaving, microwaving and mutilation shredding was carried for waste under category 3, 4, 6, 7, 8 and 10. Category 1 and 2 can be disposed off by deep burial. Category 9 waste was disposed by municipal landfill.

References

- Introductory-chapter-introduction-to-hazardous-waste-management, management-of-hazardous-wastes: intechopen.com, Retrieved 24, August 2020
- Operationalising-municipal-solid-waste-management, integrated-waste-management-volume-ii: intechopen.com, Retrieved 06, May 2020
- Biomedical-waste-and-its-management-308294194: researchgate.net, Retrieved 16, July 2020

4

Environmental Laws and Policies

The laws which seek to protect the environment are termed as environmental laws. A few types of environmental laws are air quality laws, water quality laws and waste management laws. This chapter has been carefully written to provide an easy understanding of the varied facets of environmental laws and policies such as environmental governance, environmental impact assessment and environmental globalization.

Environmental Policy

Environmental policy refers to the commitment of an organization to the laws, regulations, and other policy mechanisms concerning environmental issues. These issues generally include air and water pollution, solid waste management, ecosystem management, maintenance of biodiversity, the protection of natural resources, wildlife and endangered species. Policies concerning energy or regulation of toxic substances including pesticides and many types of industrial waste are part of the topic of environmental policy. This policy can be deliberately taken to direct and oversee human activities and thereby prevent harmful effects on the biophysical environment and natural resources, as well as to make sure that changes in the environment do not have harmful effects on humans.

Definition

It is useful to consider that environmental policy comprises two major terms: environment and policy. Environment refers to the physical ecosystems, but can also take into consideration the social dimension (quality of life, health) and an economic dimension (resource management, biodiversity). Policy can be defined as a "course of action or principle adopted or proposed by a government, party, business or individual". Thus, environmental policy focuses on problems arising from human impact on the environment, which retroacts onto human society by having a (negative) impact on human values such as good health or the 'clean and green' environment.

Environmental issues generally addressed by environmental policy include (but are not limited to) air and water pollution, waste management, ecosystem management, biodiversity protection, the protection of natural resources, wildlife and endangered species, and the preservation of these natural resources for future generations. Relatively recently, environmental policy has also attended to the communication of environmental issues.

Rationale

The rationale for governmental involvement in the environment is market failure in the form of forces beyond the control of one person, including the free rider problem and the tragedy of the commons. An example of an externality is when a factory produces waste pollution which may be dumped into a river, ultimately contaminating water. The cost of such action is paid by society-at-large, when they must clean the water before drinking it and is external to the costs of the factory. The free rider problem is when the private marginal cost of taking action to protect the environment is greater than the private marginal benefit, but the social marginal cost is less than the social marginal benefit. The tragedy of the commons is the problem that, because no one person owns the commons, each individual has an incentive to utilize common resources as much as possible. Without governmental involvement, the commons is overused. Examples of tragedies of the commons are overfishing and overgrazing.

Instruments, Problems and Issues

Environmental policy instruments are tools used by governments to implement their environmental policies. Governments may use a number of different types of instruments. For example, economic incentives and market-based instruments such as taxes and tax exemptions, tradable permits, and fees can be very effective to encourage compliance with environmental policy.

Bilateral agreements between the government and private firms and commitments made by firms independent of government requirement are examples of voluntary environmental measures. Another instrument is the implementation of greener public purchasing programs.

Several instruments are sometimes combined in a policy mix to address a certain environmental problem. Since environmental issues have many aspects, several policy instruments may be needed to adequately address each one. Furthermore, a combination of different policies may give firms greater flexibility in policy compliance and reduce uncertainty as to the cost of such compliance.

Government policies must be carefully formulated so that the individual measures do not undermine one another, or create a rigid and cost-ineffective framework. Overlapping policies result in unnecessary administrative costs, increasing the cost of implementation. To help governments realize their policy goals, the OECD Environment Directorate collects data on the efficiency and consequences of environmental policies implemented by the national governments. The website, www.economicinstruments.com, provides database detailing countries' experiences with their environmental policies. The United Nations Economic Commission for Europe, through UNECE Environmental Performance Reviews, evaluates progress made by its member countries in improving their environmental policies.

The current reliance on a market-based framework is controversial, however, and many environmentalists contend that a more radical, overarching approach is needed than a set of specific initiatives, to deal with the climate change. For example, energy efficiency measures may actually increase energy consumption in the absence of a cap on fossil fuel use, as people might drive more fuel-efficient cars. Thus, Aubrey Meyer calls for a 'framework-based market' of contraction and convergence. The Cap and Share and the Sky Trust are proposals based on the idea.

Environmental impact assessments (EIA) are conducted to compare impacts of various policy alternatives. Moreover, it is assumed that policymakers make rational decisions based on the merits of the project. Eccleston and March argue that although policymakers normally have access to reasonably accurate information, political and economic factors often lead to environmentally destructive decisions in the long run.

The decision-making theory casts doubt on this premise. Irrational decisions are reached based on unconscious biases, illogical assumptions, and the desire to avoid ambiguity and uncertainty.

Eccleston identifies and describes five of the most critical environmental policy issues facing humanity: water scarcity, food scarcity, climate change, the peak oil, and the population paradox.

Research and Innovation Policy

Synergic to the environmental policy is the environmental research and innovation policy. An example is the European environmental research and innovation policy, which aims at defining and implementing a transformative agenda to greening the economy and the society as a whole so to achieve a truly sustainable development. Europe is particularly active in this field, via a set of strategies, actions and programmes to promote more and better research and innovation for building a resource-efficient, climate resilient society and thriving economy in sync with its natural environment. Research and innovation in Europe are financially supported by the programme Horizon 2020, which is also open to participation worldwide.

History

The 1960s marked the beginning of modern environmental policy making. Although mainstream America remained oblivious to environmental concerns, the stage had been set for change by the publication of Rachel Carson's New York Times bestseller Silent Spring in 1962. Earth Day founder Gaylord Nelson, then a U.S. Senator from Wisconsin, after witnessing the ravages of the 1969 massive oil spill in Santa Barbara, California. Administrator Ruckelshaus was confirmed by the Senate on December 2, 1970, which is the traditional date used as the birth of the agency. Five months earlier, in July 1970, President Nixon had signed Reorganization Plan No. 3 calling for the establishment of EPA in July 1970. At the time, Environmental Policy was a bipartisan issue and the efforts of the United States of America helped spark countries around the world to create environmental policies. During this period, legislation was passed to regulate pollutants that go into the air, water tables, and solid waste disposal. President Nixon signed the Clean Air Act in 1970 which set the USA as one of the world leaders in environmental conservation.

In the European Union, the very first Environmental Action Programmed was adopted by national government representatives in July 1973 during the first meeting of the Council of Environmental Ministers. Since then an increasingly dense network of legislation has developed, which now extends to all areas of environmental protection including air pollution control, water protection and waste policy but also nature conservation and the control of chemicals, biotechnology and other industrial risks. EU environmental policy has thus become a core area of European politics.

Overall organizations are becoming more aware of their environmental risks and performance requirements. In line with the ISO 14001 standard they are developing environmental policies suitable for their organization. This statement outlines environmental performance of the organization as well as its environmental objectives. Written by top management of the organization they document a commitment to continuous improvement and complying with legal and other requirements, such as the environmental policy objectives set by their governments.

Environmental Policy Integration

The concept of environmental policy integration (EPI) refers to the process of integrating environmental objectives into non-environmental policy areas, such as energy, agriculture and transport, rather than leaving them to be pursued solely through purely environmental policy practices. This is oftentimes particularly challenging because of the need to reconcile global objectives and international rules with domestic needs and laws. EPI is widely recognised as one of the key elements of sustainable development. More recently, the notion of 'climate policy integration', also denoted as 'mainstreaming', has been applied to indicate the integration of climate considerations (both mitigation and adaptation) into the normal (often economically focused) activity of government.

Environmental Policy Studies

Given the growing need for trained environmental practitioners, graduate schools throughout the world offer specialized professional degrees in environmental policy studies. While there is not a standard curriculum, students typically take classes in policy analysis, environmental science, environmental law and politics, ecology, energy, and natural resource management. Graduates of these programs are employed by governments, international organizations, private sector, think tanks, universities, and so on.

Due to the lack of standard nomenclature, institutions use varying designations to refer to academic degrees they award. However, the degrees typically fall in one of four broad categories: master of arts, master of science, master of public administration, and PhD in environmental policy. Sometimes, more specific names are used to reflect the focus of the academic program. For example, the Middlebury Institute of International Studies at Monterey awards master of arts in international environmental policy (MAIEP) to emphasize the international orientation of the curriculum.

Environmental Law

Environmental law - or "environmental and natural resources law" - is a collective term describing the network of treaties, statutes, regulations, and common and customary laws addressing the effects of human activity on the natural environment.

Regulatory Subjects

The broad category of "environmental law" may be broken down into a number of more specific regulatory subjects. While there is no single agreed-upon taxonomy, the core environmental law regimes address environmental pollution. A related but distinct set of regulatory regimes, now strongly influenced by environmental legal principles, focus on the management of specific natural resources, such as forests, minerals, or fisheries. Other areas, such as environmental impact assessment, may not fit neatly into either category, but are nonetheless important components of environmental law.

Impact Assessment

Environmental impact assessment (EA) is the term used for the assessment of the environmental consequences (positive and negative) of a plan, policy, program, or project prior to the decision to move forward with the proposed action. In this context, the term 'environmental impact assessment' (EIA) is usually used when applied to concrete projects and the term 'strategic environmental assessment' applies to policies, plans and programmes (Fischer, 2016). Environmental assessments may be governed by rules of administrative procedure regarding public participation and documentation of decision making, and may be subject to judicial review.

Air Quality

Industrial air pollution now regulated by air quality law.

Air quality laws govern the emission of air pollutants into the atmosphere. A specialized subset of air quality laws regulate the quality of air inside buildings. Air quality laws are often designed specifically to protect human health by limiting or eliminating airborne pollutant concentrations. Other initiatives are designed to address broader ecological problems, such as limitations on chemicals that affect the ozone layer, and emissions trading programs to address acid rain or climate change. Regulatory efforts include identifying and categorizing air pollutants, setting limits on acceptable emissions levels, and dictating necessary or appropriate mitigation technologies.

Water Quality

A typical stormwater outfall, subject to water quality law.

Water quality laws govern the release of pollutants into water resources, including surface water, ground water, and stored drinking water. Some water quality laws, such as drinking water regulations, may be designed solely with reference to human health. Many others, including restrictions on the alteration of the chemical, physical, radiological, and biological characteristics of water resources, may also reflect efforts to protect aquatic ecosystems more broadly. Regulatory efforts may include identifying and categorizing water pollutants, dictating acceptable pollutant concentrations in water resources, and limiting pollutant discharges from effluent sources. Regulatory areas include sewage treatment and disposal, industrial and agricultural waste water management, and control of surface runoff from construction sites and urban environments.

Waste Management

A municipal landfill, operated pursuant to waste management law.

Waste management laws govern the transport, treatment, storage, and disposal of all manner of waste, including municipal solid waste, hazardous waste, and nuclear waste, among many other types. Waste laws are generally designed to minimize or eliminate the uncontrolled dispersal of waste materials into the environment in a manner that may cause ecological or biological harm, and include laws designed to reduce the generation of waste and promote or mandate waste recycling. Regulatory efforts include identifying and categorizing waste types and mandating transport, treatment, storage, and disposal practices.

Contaminant Cleanup

Oil spill emergency response, governed by environmental cleanup law.

Environmental cleanup laws govern the removal of pollution or contaminants from environmental media such as soil, sediment, surface water, or ground water. Unlike pollution control laws, cleanup laws are designed to respond after-the-fact to environmental contamination, and consequently must often define not only the necessary response actions, but also the parties who may be responsible for undertaking (or paying for) such actions. Regulatory requirements may include rules for emergency response, liability allocation, site assessment, remedial investigation, feasibility studies, remedial action, post-remedial monitoring, and site reuse.

Chemical Safety

Chemical safety laws govern the use of chemicals in human activities, particularly man-made chemicals in modern industrial applications. As contrasted with media-oriented environmental laws (e.g., air or water quality laws), chemical control laws seek to manage the (potential) pollutants themselves. Regulatory efforts include banning specific chemical constituents in consumer products (e.g., Bisphenol A in plastic bottles), and regulating pesticides.

Water Resources

An irrigation ditch, operated in accordance with water resources law.

Water resources laws govern the ownership and use of water resources, including surface water and ground water. Regulatory areas may include water conservation, use restrictions, and ownership regimes.

Mineral Resources

Mineral resource laws cover several basic topics, including the ownership of the mineral resource and who can work them. Mining is also affected by various regulations regarding the health and safety of miners, as well as the environmental impact of mining.

Forest Resources

A timber operation, regulated by forestry law.

Forestry laws govern activities in designated forest lands, most commonly with respect to forest management and timber harvesting. Ancillary laws may regulate forest land acquisition and prescribed burn practices. Forest management laws generally adopt management policies, such as multiple use and sustained yield, by which public forest resources are to be managed. Governmental agencies are generally responsible for planning and implementing forestry laws on public forest lands, and may be involved in forest inventory, planning, and conservation, and oversight of timber sales. Broader initiatives may seek to slow or reverse deforestation.

Wildlife and Plants

Wildlife laws govern the potential impact of human activity on wild animals, whether directly on individuals or populations, or indirectly via habitat degradation. Similar laws may operate to protect plant species. Such laws may be enacted entirely to protect biodiversity, or as a means for protecting species deemed important for other reasons. Regulatory efforts may including the creation of special conservation statuses, prohibitions on killing, harming, or disturbing protected species, efforts to induce and support species recovery, establishment of wildlife refuges to support conservation, and prohibitions on trafficking in species or animal parts to combat poaching.

Fish and Game

Fish and game laws regulate the right to pursue and take or kill certain kinds of fish and wild animal (game). Such laws may restrict the days to harvest fish or game, the number of animals caught per person, the species harvested, or the weapons or fishing gear used. Such laws may seek to balance dueling needs for preservation and harvest and to manage both environment and populations of fish and game. Game laws can provide a legal structure to collect license fees and other money which is used to fund conservation efforts as well as to obtain harvest information used in wildlife management practice.

Important Principles

Environmental law has developed in response to emerging awareness of and concern over issues impacting the entire world. While laws have developed piecemeal and for a variety of reasons, some effort has gone into identifying key concepts and guiding principles common to environmental law as a whole. The principles discussed below are not an exhaustive list and are not universally recognized or accepted. Nonetheless, they represent important principles for the understanding of environmental law around the world.

Sustainable Development

Defined by the United Nations Environment Programme as “development that meets the needs of the present without compromising the ability of future generations to meet their own needs,” sustainable development may be considered together with the concepts of “integration” (development cannot be considered in isolation from sustainability) and “interdependence” (social and economic development, and environmental protection, are interdependent). Laws mandating environmental impact assessment and requiring or encouraging development to minimize environmental impacts may be assessed against this principle.

The modern concept of sustainable development was a topic of discussion at the 1972 United Nations Conference on the Human Environment (Stockholm Conference), and the driving force behind the 1983 World Commission on Environment and Development (WCED, or Bruntland Commission). In 1992, the first UN Earth Summit resulted in the Rio Declaration, Principle 3 of which reads: "The right to development must be fulfilled so as to equitably meet developmental and environmental needs of present and future generations." Sustainable development has been a core concept of international environmental discussion ever since, including at the World Summit on Sustainable Development (Earth Summit 2002), and the United Nations Conference on Sustainable Development (Earth Summit 2012, or Rio+20).

Equity

Defined by UNEP to include intergenerational equity - "the right of future generations to enjoy a fair level of the common patrimony" - and intragenerational equity - "the right of all people within the current generation to fair access to the current generation's entitlement to the Earth's natural resources" - environmental equity considers the present generation under an obligation to account for long-term impacts of activities, and to act to sustain the global environment and resource base for future generations. Pollution control and resource management laws may be assessed against this principle.

Transboundary Responsibility

Defined in the international law context as an obligation to protect one's own environment, and to prevent damage to neighboring environments, UNEP considers transboundary responsibility at the international level as a potential limitation on the rights of the sovereign state. Laws that act to limit externalities imposed upon human health and the environment may be assessed against this principle.

Public Participation and Transparency

Identified as essential conditions for "accountable governments . . ., industrial concerns," and organizations generally, public participation and transparency are presented by UNEP as requiring "effective protection of the human right to hold and express opinions and to seek, receive and impart ideas," "a right of access to appropriate, comprehensible and timely information held by governments and industrial concerns on economic and social policies regarding the sustainable use of natural resources and the protection of the environment, without imposing undue financial burdens upon the applicants and with adequate protection of privacy and business confidentiality," and "effective judicial and administrative proceedings." These principles are present in environmental impact assessment, laws requiring publication and access to relevant environmental data, and administrative procedure.

Precautionary Principle

One of the most commonly encountered and controversial principles of environmental law, the Rio Declaration formulated the precautionary principle as follows:

> In order to protect the environment, the precautionary approach shall be widely applied

> by States according to their capabilities. Where there are threats of serious or irreversible damage, lack of full scientific certainty shall not be used as a reason for postponing cost-effective measures to prevent environmental degradation.

The principle may play a role in any debate over the need for environmental regulation.

Prevention

> The concept of prevention . . . can perhaps better be considered an overarching aim that gives rise to a multitude of legal mechanisms, including prior assessment of environmental harm, licensing or authorization that set out the conditions for operation and the consequences for violation of the conditions, as well as the adoption of strategies and policies. Emission limits and other product or process standards, the use of best available techniques and similar techniques can all be seen as applications of the concept of prevention.

Polluter Pays Principle

The polluter pays principle stands for the idea that "the environmental costs of economic activities, including the cost of preventing potential harm, should be internalized rather than imposed upon society at large." All issues related to responsibility for cost for environmental remediation and compliance with pollution control regulations involve this principle.

History

Early examples of legal enactments designed to consciously preserve the environment, for its own sake or human enjoyment, are found throughout history. In the common law, the primary protection was found in the law of nuisance, but this only allowed for private actions for damages or injunctions if there was harm to land. Thus smells emanating from pig stys, strict liability against dumping rubbish, or damage from exploding dams. Private enforcement, however, was limited and found to be woefully inadequate to deal with major environmental threats, particularly threats to common resources. During the "Great Stink" of 1858, the dumping of sewerage into the River Thames began to smell so ghastly in the summer heat that Parliament had to be evacuated. Ironically, the Metropolitan Commission of Sewers Act 1848 had allowed the Metropolitan Commission for Sewers to close cesspits around the city in an attempt to "clean up" but this simply led people to pollute the river. In 19 days, Parliament passed a further Act to build the London sewerage system. London also suffered from terrible air pollution, and this culminated in the "Great Smog" of 1952, which in turn triggered its on legislative response: the Clean Air Act 1956. The basic regulatory structure was to set limits on emissions for households and business (particularly burning coal) while an inspectorate would enforce compliance.

Notwithstanding early analogues, the concept of "environmental law" as a separate and distinct body of law is a twentieth-century development. The recognition that the natural environment was fragile and in need of special legal protections, the translation of that recognition into legal structures, the development of those structures into a larger body of "environmental law," and the strong influence of environmental law on natural resource laws, did not occur until about the

1960s. At that time, numerous influences - including a growing awareness of the unity and fragility of the biosphere; increased public concern over the impact of industrial activity on natural resources and human health; the increasing strength of the regulatory state; and more broadly the advent and success of environmentalism as a political movement - coalesced to produce a huge new body of law in a relatively short period of time. While the modern history of environmental law is one of continuing controversy, by the end of the twentieth century environmental law had been established as a component of the legal landscape in all developed nations of the world, many developing ones, and the larger project of international law.

Controversy

Environmental law is a continuing source of controversy. Debates over the necessity, fairness, and cost of environmental regulation are ongoing, as well as regarding the appropriateness of regulations vs. market solutions to achieve even agreed-upon ends.

Allegations of scientific uncertainty fuel the ongoing debate over greenhouse gas regulation, and are a major factor in debates over whether to ban particular pesticides. In cases where the science is well-settled, it is not unusual to find that corporations intentionally hide or distort the facts, or sow confusion.

It is very common for regulated industry to argue against environmental regulation on the basis of cost. Difficulties arise in performing cost-benefit analysis of environmental issues. It is difficult to quantify the value of an environmental value such as a healthy ecosystem, clean air, or species diversity. Many environmentalists' response to pitting economy vs. ecology is summed up by former Senator and founder of Earth Day Gaylord Nelson, "The economy is a wholly owned subsidiary of the environment, not the other way around." Furthermore, environmental issues are seen by many as having an ethical or moral dimension, which would transcend financial cost. Even so, there are some efforts underway to systemically recognize environmental costs and assets, and account for them properly in economic terms.

While affected industries spark controversy in fighting regulation, there are also many environmentalists and public interest groups who believe that current regulations are inadequate, and advocate for stronger protection. Environmental law conferences - such as the annual Public Interest Environmental Law Conference in Eugene, Oregon - typically have this focus, also connecting environmental law with class, race, and other issues.

Around the World

International Law

Global and regional environmental issues are increasingly the subject of international law. Debates over environmental concerns implicate core principles of international law and have been the subject of numerous international agreements and declarations.

Customary international law is an important source of international environmental law. These are the norms and rules that countries follow as a matter of custom and they are so prevalent that they bind all states in the world. When a principle becomes customary law is not clear cut and many arguments are put forward by states not wishing to be bound. Examples of customary international law relevant to the environment include the duty to warn other states promptly about icons of an environmental nature and environmental damages to which another state or states may be exposed, and Principle 21 of the Stockholm Declaration ('good neighbourliness' or sic utere).

Numerous legally binding international agreements encompass a wide variety of issue-areas, from terrestrial, marine and atmospheric pollution through to wildlife and biodiversity protection. International environmental agreements are generally multilateral (or sometimes bilateral) treaties (a.k.a. convention, agreement, protocol, etc.). Protocols are subsidiary agreements built from a primary treaty. They exist in many areas of international law but are especially useful in the environmental field, where they may be used to regularly incorporate recent scientific knowledge. They also permit countries to reach agreement on a framework that would be contentious if every detail were to be agreed upon in advance. The most widely known protocol in international environmental law is the Kyoto Protocol, which followed from the United Nations Framework Convention on Climate Change.

While the bodies that proposed, argued, agreed upon and ultimately adopted existing international agreements vary according to each agreement, certain conferences, including 1972's United Nations Conference on the Human Environment, 1983's World Commission on Environment and Development, 1992's United Nations Conference on Environment and Development and 2002's World Summit on Sustainable Development have been particularly important. Multilateral environmental agreements sometimes create an International Organization, Institution or Body responsible for implementing the agreement. Major examples are the Convention on International Trade in Endangered Species of Wild Fauna and Flora (CITES) and the International Union for Conservation of Nature (IUCN).

International environmental law also includes the opinions of international courts and tribunals. While there are few and they have limited authority, the decisions carry much weight with legal commentators and are quite influential on the development of international environmental law. One of the biggest challenges in international decisions is to determine an adequate compensation for environmental damages. The courts include the International Court of Justice (ICJ); the international Tribunal for the Law of the Sea (ITLOS); the European Court of Justice; European Court of Human Rights and other regional treaty tribunals.

Africa

According to the International Network for Environmental Compliance and Enforcement (INECE), the major environmental issues in Africa are "drought and flooding, air pollution, deforestation, loss of biodiversity, freshwater availability, degradation of soil and vegetation, and widespread poverty." The U.S. Environmental Protection Agency (EPA) is focused on the "growing urban and industrial pollution, water quality, electronic waste and indoor air from cookstoves." They hope to provide enough aid on concerns regarding pollution before their impacts contaminate the African environment as well as the global environment. By doing so, they intend to "protect human health, particularly vulnerable populations such as children and the poor." In order to accomplish these goals in Africa, EPA programs are focused on strengthening the ability to enforce environmental laws as well as public compliance to them. Other programs work on developing stronger environmental laws, regulations, and standards.

Asia

The Asian Environmental Compliance and Enforcement Network (AECEN) is an agreement between 16 Asian countries dedicated to improving cooperation with environmental laws in Asia. These countries include Cambodia, China, Indonesia, India, Maldives, Japan, Korea, Malaysia, Nepal, Philippines, Pakistan, Singapore, Sri Lanka, Thailand, Vietnam, and Lao PDR.

European Union

The European Union issues secondary legislation on environmental issues that are valid throughout the EU (so called regulations) and many directives that must be implemented into national legislation from the 28 member states (national states). Examples are the Regulation (EC) No. 338/97 on the implementation of CITES; or the Natura 2000 network the centerpiece for nature & biodiversity policy, encompassing the bird Directive (79/409/EEC/ changed to 2009/147/EC) and the habitats directive (92/43/EEC). Which are made up of multiple SACs (Special Areas of Conservation, linked to the habitats directive) & SPAs (Special Protected Areas, linked to the bird directive), throughout Europe.

EU legislation is ruled in Article 249 Treaty for the Functioning of the European Union (TFEU). Topics for common EU legislation are:

- Climate change.
- Air pollution.
- Water protection and management.
- Waste management.
- Soil protection.
- Protection of nature, species and biodiversity.
- Noise pollution.
- Cooperation for the environment with third countries (other than EU member states).
- Civil protection.

Middle East

The U.S. Environmental Protection Agency is working with countries in the Middle East to improve "environmental governance, water pollution and water security, clean fuels and vehicles, public participation, and pollution prevention."

Ocean

The main concerns on environmental issues in the Oceanic Region are "illegal releases of air and water pollutants, illegal logging/timber trade, illegal shipment of hazardous wastes, including e-waste and ships slated for destruction, and insufficient institutional structure/lack of enforcement capacity". The Secretariat of the Pacific Regional Environmental Programme (SPREP) is an international organization between Australia, the Cook Islands, FMS, Fiji, France, Kiribati, Marshall Islands, Nauru, New Zealand, Niue, Palau, PNG, Samoa, Solomon Island, Tonga, Tuvalu, USA, and Vanuatu. The SPREP was established in order to provide assistance in improving and protecting the environment as well as assure sustainable development for future generations.

Australia

The Environment Protection and Biodiversity Conservation Act 1999 is the center piece of environmental legislation in the Australian Government. It sets up the "legal framework to protect and manage nationally and internationally important flora, fauna, ecological communities and heritage places". It also focuses on protecting world heritage properties, national heritage properties, wetlands of international importance, nationally threatened species and ecological communities, migratory species, Commonwealth marine areas, Great Barrier Reef Marine Park, and the environment surrounding nuclear activities. *Commonwealth v Tasmania* (1983), also known as the "Tasmanian Dam Case", is the most influential case for Australian environmental law.

Brazil

The Brazilian government created the Ministry of Environment in 1992 in order to develop better strategies of protecting the environment, use natural resources sustainably, and enforce public environmental policies. The Ministry of Environment has authority over policies involving environment, water resources, preservation, and environmental programs involving the Amazon.

Canada

The Department of the Environment Act establishes the Department of the Environment in the Canadian government as well as the position Minister of the Environment. Their duties include "the preservation and enhancement of the quality of the natural environment, including water, air and soil quality; renewable resources, including migratory birds and other non-domestic flora and fauna; water; meteorology;" The Environmental Protection Act is the main piece of Canadian environmental legislation that was put into place March 31, 2000. The Act focuses on "respecting pollution prevention and the protection of the environment and human health in order to contribute to sustainable development." Other principle federal statutes include the Canadian Environmental Assessment Act, and the Species at Risk Act. When provincial and federal legislation are in conflict federal legislation takes precedence, that being said individual provinces can have their own legislation such as Ontario's Environmental Bill of Rights, and Clean Water Act.

China

According to the U.S. Environmental Protection Agency, "China has been working with great determination in recent years to develop, implement, and enforce a solid environmental law framework. Chinese officials face critical challenges in effectively implementing the laws, clarifying the roles of their national and provincial governments, and strengthening the operation of their legal system." Explosive economic and industrial growth in China has led to significant environmental degradation, and China is currently in the process of developing more stringent legal controls. The harmonization of Chinese society and the natural environment is billed as a rising policy priority.

Ecuador

With the enactment of the 2008 Constitution, Ecuador became the first country in the world to codify the Rights of Nature. The Constitution, specifically Articles 10 and 71-74, recognizes the inalienable rights of ecosystems to exist and flourish, gives people the authority to petition on the behalf of ecosystems, and requires the government to remedy violations of these rights. The rights approach is a break away from traditional environmental regulatory systems, which regard nature as property and legalize and manage degradation of the environment rather than prevent it.

The Rights of Nature articles in Ecuador's constitution are part of a reaction to a combination of political, economic, and social phenomena. Ecuador's abusive past with the oil industry, most famously the class-action litigation against Chevron, and the failure of an extraction-based economy and neoliberal reforms to bring economic prosperity to the region has resulted in the election of a New Leftist regime, led by President Rafael Correa, and sparked a demand for new approaches to development. In conjunction with this need, the principle of "Buen Vivir," or good living—focused on social, environmental and spiritual wealth versus material wealth—gained popularity among citizens and was incorporated into the new constitution.

The influence of indigenous groups, from whom the concept of "Buen Vivir" originates, in the forming of the constitutional ideals also facilitated the incorporation of the Rights of Nature as a basic tenet of their culture and conceptualization of "Buen Vivir."

Egypt

The Environmental Protection Law outlines the responsibilities of the Egyptian government to "preparation of draft legislation and decrees pertinent to environmental management, collection of data both nationally and internationally on the state of the environment, preparation of periodical reports and studies on the state of the environment, formulation of the national plan and its projects, preparation of environmental profiles for new and urban areas, and setting of standards to be used in planning for their development, and preparation of an annual report on the state of the environment to be prepared to the President."

India

In India, Environmental law is governed by the Environment Protection Act, 1986. This act is enforced by the Central Pollution Control Board and the numerous State Pollution Control Boards. Apart from this, there are also individual legislations specifically enacted for the protection of Water, Air, Wildlife, etc. Such legislations include :-

- The Water (Prevention and Control of Pollution) Act, 1974.
- The Water (Prevention and Control of Pollution) Cess Act, 1977.
- The Forest (Conservation) Act, 1980.
- The Air (Prevention and Control of Pollution) Act, 1981.
- Air (Prevention and Control of Pollution) (Union Territories) Rules, 1983.
- The Biological Diversity Act, 2002 and the Wild Life Protection Act, 1972.

- Batteries (Management and Handling) Rules, 2001.
- Recycled Plastics, Plastics Manufacture and Usage Rules, 1999.
- The National Green Tribunal established under the National Green Tribunal Act of 2010 has jurisdiction over all environmental cases dealing with a substantial environmental question and acts covered under the Water (Prevention and Control of Pollution) Act, 1974.
- Water (Prevention and Control of Pollution) Cess Rules, 1978.
- Ganga Action Plan, 1986.
- The Forest (Conservation) Act, 1980.
- The Public Liability Insurance Act, 1991 and the Biological Diversity Act, 2002. The acts covered under Indian Wild Life Protection Act 1972 do not fall within the jurisdiction of the National Green Tribunal. Appeals can be filed in the Hon'ble Supreme Court of India.
- Basel Convention on Control of TransboundaryMovements on Hazardous Wastes and Their Disposal, 1989 and Its Protocols.
- Hazardous Wastes (Management and Handling) Amendment Rules, 2003.

Japan

The Basic Environmental Law is the basic structure of Japan's environmental policies replacing the Basic Law for Environmental Pollution Control and the Nature Conservation Law. The updated law aims to address "global environmental problems, urban pollution by everyday life, loss of accessible natural environment in urban areas and degrading environmental protection capacity in forests and farmlands."

The three basic environmental principles that the Basic Environmental Law follows are "the blessings of the environment should be enjoyed by the present generation and succeeded to the future generations, a sustainable society should be created where environmental loads by human activities are minimized, and Japan should contribute actively to global environmental conservation through international cooperation." From these principles, the Japanese government have established policies such as "environmental consideration in policy formulation, establishment of the Basic Environment Plan which describes the directions of long-term environmental policy, environmental impact assessment for development projects, economic measures to encourage activities for reducing environmental load, improvement of social infrastructure such as sewerage system, transport facilities etc., promotion of environmental activities by corporations, citizens and NGOs, environmental education, and provision of information, promotion of science and technology."

New Zealand

The Ministry for the Environment and Office of the Parliamentary Commissioner for the Environment were established by the Environment Act 1986. These positions are responsible for advising the Minister on all areas of environmental legislation. A common theme of New Zealand's environmental legislation is sustainably managing natural and physical resources, fisheries, and forests. The Resource Management Act 1991 is the main piece of environmental legislation that outlines the government's strategy to managing the "environment, including air, water soil, biodiversity, the coastal environment, noise, subdivision, and land use planning in general."

Russia

The Ministry of Natural Resources and Environment of the Russian Federation makes regulation regarding "conservation of natural resources, including the subsoil, water bodies, forests located in designated conservation areas, fauna and their habitat, in the field of hunting, hydrometeorology and related areas, environmental monitoring and pollution control, including radiation monitoring and control, and functions of public environmental policy making and implementation and statutory regulation."

Vietnam

Vietnam is currently working with the U.S. Environmental Protection Agency on dioxin remediation and technical assistance in order to lower methane emissions. In March 2002, the U.S and Vietnam signed the U.S.-Vietnam Memorandum of Understanding on Research on Human Health and the Environmental Effects of Agent Orange/Dioxin.

Environmental Justice

Environmental justice emerged as a concept in the United States in the early 1980s. The term has two distinct uses. The first and more common usage describes a social movement whose focus is on the fair distribution of environmental benefits and burdens. Second, it is an interdisciplinary body of social science literature that includes theories of the environment, theories of justice, environmental law and governance, environmental policy and planning, development, sustainability, and political ecology.

Definition

The United States Environmental Protection Agency defines environmental justice as follows:

Environmental justice is the fair treatment and meaningful involvement of all people regardless of race, color, national origin, or income with respect to the development, implementation, and enforcement of environmental laws, regulations, and policies. EPA has this goal for all communities and persons across this Nation [*sic*]. It will be achieved when everyone enjoys the same degree of protection from environmental and health hazards and equal access to the decision-making process to have a healthy environment in which to live, learn, and work.

Other definitions include equitable distribution of environmental risks and benefits; fair and meaningful participation in environmental decision-making; recognition of community ways of life, local knowledge, and cultural difference; and the capability of communities and individuals to function and flourish in society.

Environmental Discrimination

One issue that environmental justice seeks to address is that of environmental discrimination. Racism and discrimination against minorities center on a socially-dominant group's belief in its superiority, often resulting in a) privilege for the dominant group and b) the mistreatment of non-dominant minorities. The combined impact of these privileges and prejudices are just one of the potential reasons that waste management and highly-pollutive sites tend to be located in minority-dominated areas. A disproportionate quantity of minority communities (for example in Warren County, North Carolina) play host to landfills, incinerators, and other potentially toxic facilities.

Environmental discrimination has historically been evident in the process of selecting and building environmentally hazardous sites, including waste disposal, manufacturing, and energy production facilities. The location of transportation infrastructures, including highways, ports, and airports, has also been viewed as a source of environmental injustice. Among the earliest documentation of environmental racism was a study of the distribution of toxic waste sites across the United States. Due to the results of that study, waste dumps and waste incinerators have been the target of environmental justice lawsuits and protests.

Litigation

Some environmental justice lawsuits are based on violations of civil rights laws.

Title VI of the Civil Rights Act of 1964 is often used in lawsuits that claim environmental inequality. Section 601 prohibits discrimination based on race, color, or national origin by any government agency receiving federal assistance. To win an environmental justice case that claims an agency violated this statute, the plaintiff must prove the agency intended to discriminate. Section 602 requires agencies to create rules and regulations that uphold section 601. This section is useful because the plaintiff must only prove that the rule or regulation in question had a discriminatory impact. There is no need to prove discriminatory intent. *Seif v. Chester Residents Concerned for Quality Living* set the precedent that citizens can sue under section 601. There has not yet been a case in which a citizen has sued under section 602, which calls into question whether this right of action exists.

The Equal Protection Clause of the Fourteenth Amendment, which was used many times to defend minority rights during the 1960s, has also been used in numerous environmental justice cases.

Initial Barriers to Minority Participation

When environmentalism first became popular during the first half of the 20th century, the focus was wilderness protection and wildlife preservation. These goals reflected the interests of the movement's initial supporters. The actions of many mainstream environmental organizations still reflect these early principles.

Many low-income minorities felt isolated or even negatively impacted by the movement, exemplified by the Southwest Organizing Project's (SWOP) Letter to the Group of 10, a letter sent to major environmental organizations by several local environmental justice activists. The letter argued that the environmental movement was so concerned about cleaning up and preserving nature that it ignored the negative side-effects that doing so caused communities nearby, namely less job growth. In addition, the NIMBY movement has transferred locally unwanted land uses (LULUs) from middle-class neighborhoods to poor communities with large minority populations. Therefore, vulnerable communities with fewer political opportunities are more often exposed to hazardous waste and toxins. This has resulted in the PIBBY principle, or at least the PIMBY (Place-in-minorities'-backyard), as supported by the United Church of Christ's study in 1987.

As a result, some minorities have viewed the environmental movement as elitist. Environmental elitism manifested itself in three different forms:

1. *Compositional* – Environmentalists are from the middle and upper class.
2. *Ideological* – The reforms benefit the movement's supporters but impose costs on nonparticipants.
3. *Impact* – The reforms have "regressive social impacts". They disproportionately benefit environmentalists and harm underrepresented populations.

Supporters of economic growth have taken advantage of environmentalists' neglect of minorities. They have convinced minority leaders looking to improve their communities that the economic benefits of industrial facility and the increase in the number of jobs are worth the health risks. In fact, both politicians and businesses have even threatened imminent job loss if communities do not accept hazardous industries and facilities. Although in many cases local residents do not actually receive these benefits, the argument is used to decrease resistance in the communities as well as avoid expenditures used to clean up pollutants and create safer workplace environments.

Cost Barriers

One of the major initial barriers to minority participation in environmental justice is the initial costs of trying to change the system and prevent companies from dumping their toxic waste and other pollutants in areas with high numbers of minorities living in them. There are massive legal fees involved in fighting for environmental justice and trying to shed environmental racism. For example, in the United Kingdom, there is a rule that the claimant may have to cover the fees of their opponents, which further exacerbates any cost issues, especially with lower income minority groups; also, the only way for environmental justice groups to hold companies accountable for their pollution and breaking any licensing issues over waste disposal would be to sue the government for not enforcing rules. This would lead to the forbidding legal fees that most could not afford. This can be seen by the fact that out of 210 judicial review cases between 2005 and 2009, 56% did not proceed due to costs.

Contributions of the Civil Rights Movement

During the African-American Civil Rights Movement in the 1960s, activists participated in a social movement that created a unified atmosphere and advocated goals of social justice and equality. The community organization and the social values of the era have translated to the Environmental Justice movement.

Similar Goals and Tactics

The Environmental Justice movement and the Civil Rights Movement have many commonalities. At their core, the goals of movements are the same: "social justice, equal protection, and an end to institutional discrimination." By stressing the similarities of the two movements, it emphasizes that environmental equity is a right for all citizens. Because the two movements have parallel goals, it is useful to employ similar tactics that often emerge on the grassroots level. Common confrontational strategies include protests, neighborhood demonstrations, picketing, political pressure, and demonstration.

Existing Organizations and Leaders

Just as the civil rights movement of the 1960s began in the South, the modern civil rights movement and the fight for environmental equity has been largely based in the South, where environmental discrimination is most prominent. In these southern communities, black churches and other voluntary associations are used to organize resistance efforts, including research and demonstrations, such as the protest in Warren County, North Carolina. As a result of the existing community structure, many church leaders and civil rights activists, such as Reverend Benjamin Chavis Muhammad, have spearheaded the Environmental Justice movement.

The Bronx, in New York city, has become a recent example of Environmental Justice succeeding. Majora Carter spearheaded the South Bronx Greenway Project, bringing local economic development, local urban heat island mitigation, positive social influences, access to public open space, and aesthetically stimulating environments. The New York City Department of Design and Construction has recently recognized the value of the South Bronx Greenway design, and consequently utilized it as a widely distributed smart growth template. This venture is the ideal shovel-ready project with over $50 million in funding.

Litigation

Some of the most successful Environmental Justice lawsuits are based on violations of civil rights laws. The first case to use civil rights as a means to legally challenge the siting of a waste facility was in 1979. With the legal representation of Linda McKeever Bullard, the wife of Robert D. Bullard, residents of Houston's Northwood Manor opposed the decision of the city and Browning Ferris Industries to construct a solid waste facility near their mostly African-American neighborhood.

In 1979, Northeast Community Action Group or NECAG, was formed by African American homeowners in a suburban, middle income neighborhood in order to keep a landfill out of their home town. This group was the first organization that found the connection between race and pollution. The group, alongside their attorney Linda McKeever Bullard started the lawsuit Bean v. Southwestern Waste Management, Inc., which was the first of its kind to challenge the sitting of a waste facility under civil rights law. The Equal Protection Clause of the Fourteenth Amendment, which was used many times to defend minority rights during the 1960s, has also been used in numerous Environmental Justice cases.

Title VI of the Civil Rights Act of 1964 is often used in lawsuits that claim environmental inequality. The two most important sections in these cases are sections 601 and 602. Section 601 prohibits discrimination based on race, color, or national origin by any government agency receiving federal assistance. To win an Environmental Justice case that claims an agency violated this statute, the plaintiff must prove the agency intended to discriminate. Section 602 requires agencies to create rules and regulations that uphold section 601. This section is useful because the plaintiff must only prove that the rule or regulation in question had a discriminatory impact. There is no need to prove discriminatory intent. *Seif v. Chester Residents Concerned for Quality Living* set the precedent that citizens can sue under section 601, there has not been a case in which a citizen has sued under section 602, which calls into question whether this right of action exists.

Affected Groups

Among the affected groups of Environmental Justice, those in high-poverty and racial minority groups have the most propensity to receive the harm of environmental injustice. Poor people account for more than 20% of the human health impacts from industrial toxic air releases, compared to 12.9% of the population nationwide. This does not account for the inequity found among individual minority groups. Some studies that test statistically for effects of race and ethnicity, while controlling for income and other factors, suggest racial gaps in exposure that persist across all bands of income.

African-Americans are affected by a variety of Environmental Justice issues. One notorious example is the "Cancer Alley" region of Louisiana. This 85-mile stretch of the Mississippi River between Baton Rouge and New Orleans is home to 125 companies that produce one quarter of the petrochemical products manufactured in the United States. The United States Commission on Civil Rights has concluded that the African-American community has been disproportionately affected by Cancer Alley as a result of Louisiana's current state and local permit system for hazardous facilities, as well as their low socio-economic status and limited political influence.

Indigenous groups are often the victims of environmental injustices. Native Americans have suffered abuses related to uranium mining in the American West. Churchrock, New Mexico, in Navajo territory was home to the longest continuous uranium mining in any Navajo land. From 1954 until 1968, the tribe leased land to mining companies who did not obtain consent from Navajo families or report any consequences of their activities. Not only did the miners significantly deplete the limited water supply, but they also contaminated what was left of the Navajo water supply with uranium. Kerr-McGee and United Nuclear Corporation, the two largest mining companies, argued that the Federal Water Pollution Control Act did not apply to them, and maintained that Native American land is not subject to environmental protections. The courts did not force them to comply with US clean water regulations until 1980.

The most common example of environmental injustice among Latinos is the exposure to pesticides faced by farmworkers. After DDT and other chlorinated hydrocarbon pesticides were banned in the United States in 1972, farmers began using more acutely toxic organophosphate pesticides such as parathion. A large portion of farmworkers in the US are working illegally, and as a result of their political disadvantage, are not able to protest against regular exposure to pesticides. Exposure to chemical pesticides in the cotton industry also affects farmers in India and Uzbekistan. Banned throughout much of the rest of the world because of the potential threat to human health and the natural environment, Endosulfan is a highly toxic chemical, the safe use of which cannot be guaranteed in the many developing countries it is used in. Endosulfan, like DDT, is an organochlorine and persists in the environment long after it has killed the target pests, leaving a deadly legacy for people and wildlife.

Residents of cities along the US-Mexico border are also affected. Maquiladoras are assembly plants operated by American, Japanese, and other foreign countries, located along the US-Mexico border. The maquiladoras use cheap Mexican labor to assemble imported components and raw material, and then transport finished products back to the United States. Much of the waste ends up being illegally dumped in sewers, ditches, or in the desert. Along the Lower Rio Grande Valley, maquiladoras dump their toxic wastes into the river from which 95 percent of residents obtain their drinking water. In the border cities of Brownsville, Texas and Matamoros, Mexico, the rate of anencephaly (babies born without brains) is four times the national average.

One reason for toxic industries to concentrate in minority neighborhoods or poor neighborhoods is because of their lack of political power. Whether it be lack of homeownership or just because of a general inability to participate politically, these groups are treated unfairly. This lack of political participation could indicate why latinos are the most affected by environmental injustice in the US, since many latinos are illegal immigrants and thus cannot participate in the political system.

States may also see placing toxic facilities near poor neighborhoods as beneficial from a Cost Benefit Analysis (CBA) perspective. Viewing a state's wealth through the lens of CBA's, it would be more favorable to place a toxic facility near a city of 20,000 poor people than it would be to place it by a city of 5,000 wealthy people. Terry Bossert of Range Resources reportedly has said that it deliberately locates its operations in poor neighbourhoods instead of wealthy areas where residents have more money to challenge its practices.

Steel works, blast furnaces, rolling and finishing mills, along with iron and steel foundries, are responsible for more than 57% of the total human health risks from industrial pollution. This means that if the government wanted to make major reformative legislation for Environmental Justice, they could easily do so by targeting these industries.

Government Agencies

U.S. Department of Agriculture

In its 2012 environmental justice strategy documents, the U.S. Department of Agriculture (USDA) stated an ongoing desire to integrate environmental justice into its core mission, internal operations and programming. It identified ambitious timeframes for action and promised improved efforts to highlight, track and coordinate EJ activities among its many sub-agencies. Agency-wide the USDA expanded its perspective on EJ, so that in addition to preventing disproportionate environmental impacts on EJ communities, USDA voiced a commitment to improve public participation processes and use its technical and financial assistance programs to improve the quality of life in all communities. In 2011, Secretary of Agriculture Tom Vilsack emphasized the USDA's focus on EJ in rural communities around the United States. USDA funds or implements many creative programs with social and environmental equity goals, however it has no staff dedicated solely to EJ, and faces the challenges of limited budgets and coordinating the efforts of a highly diverse agency.

Background

The USDA is the executive agency responsible for federal policy on food, agriculture, natural resources, and quality of life in rural America. The USDA has more than 100,000 employees and delivers over $96.5 billion in public services to programs worldwide. To fulfill its general mandate, USDA's departments are organized into seven mission areas:1) Farm and Foreign Agricultural Services; 2) Food, Nutrition and Consumer Services; 3) Food Safety; 4) Marketing and Regulatory Programs; 5) Natural Resources and Environment; 6) Research, Education and Economics and; 7) Rural Development.

In 1994, President Clinton issued Executive Order 12898, "Federal Actions to Address Environmental Justice in Minority Populations and Low-Income Populations." Executive Order 12898 requires that achieving EJ must be part of each federal agency's mission. Agency programs, policies and activities can lead to health and environmental effects that disproportionately impact minority and low-income populations. Under Executive Order 12898 agencies must develop strategies that identify and address these effects by:

1. promoting enforcement of all health and environmental statutes in areas with minority and low-income populations.
2. ensuring greater public participation.
3. improving research and data collection relating to the health and environment of minority and low-income populations.
4. identifying differential patterns of consumption of natural resources among minority and low-income populations.

Title VI of the Civil Rights Act of 1964 requires that federal funds be used in a fair and equitable manner. Under Title VI any federal agency that receives federal funding cannot discriminate. Title VI also forbids federal agencies from providing grants or funding opportunities to programs that discriminate. An agency that violates Title VI can lose its federal funding.

Following E.O. 12898 and USDA's initial EJ strategic plan, USDA issued its internal Environmental Justice Department Regulation (DR 5600-002) in 1997. Although the definition of EJ was undergoing updates in 2012, DR 5600-002 defines environmental justice as "to the greatest extent practicable and permitted by law, all populations are provided the opportunity to comment before decisions are rendered on, are allowed to share in the benefits of, are not excluded from, and are not affected in a disproportionately high and adverse manner by, government programs and activities affecting human health or the environment." Patrick Holmes, Special Assistant to the Under Secretary for Natural Resources and Environment at USDA, notes that this definition will be broadened in 2012 so that EJ also includes efforts to improve quality of life in all communities. In other words, USDA will consider EJ to include avoiding adverse impacts *and* ensuring access to environmental benefits. Further, DR 5600-002 identified USDA's goals in implementing Executive Order 12898 as:

- To incorporate environmental justice considerations into USDA's programs and activities and to address environmental justice across mission areas.
- To identify, prevent, and/or mitigate, to the greatest extent practicable, disproportionately high and adverse human health or environmental effects of USDA programs and activities on minority and low-income populations.

- To provide, to the greatest extent practicable, the opportunity for minority and low-income populations to participate in planning, analysis, and decisionmaking that affects their health or environment, including identification of program needs and designs.

DR 5600-002 is "intended only to improve the internal management of USDA," and although it described concrete, mandatory actions by the agency, it did not establish new rights or benefits enforceable in court. In April 2011, USDA Secretary Tom Vilsack has stated a more concrete priority to fulfill its mission of environmental justice in rural areas.

2012 Environmental Justice Strategy

In compliance with the August 2011 Memorandum of Understanding on Environmental Justice and Executive Order 12898 (MOU), USDA released a final Environmental Justice Strategic Plan: 2012 to 2014 on February 7, 2012 (Strategic Plan), which identifies new and updated goals and performance measures beyond what USDA identified in a 1995 EJ strategy it adopted in response to E.O. 12898. In the same week, it also released its first annual implementation progress report (Progress Report), as the MOU also required. The Secretary's message accompanying the Strategic Plan described two immediate tasks: 1) each agency within USDA is required to identify a point of contact for EJ issues, at the Senior Executive Service (SES) level; and 2) each agency must develop its own EJ strategy prior to April 15, 2012, and begin implementing it as soon as possible. As of May 2012, it did not appear that such strategies had been made public, although sub-agencies provided internal reports to the USDA's EJ steering committee on April 9, 2012, according to Holmes. The Secretary's message contained strong language that, "Given that USDA programs touch almost every American every day, the Department is well positioned to help in [the environmental justice] effort." USDA has determined that it can achieve the requirements of the Executive Order by integrating EJ into its programs, rather than implementing new and costly programs. The agency took this same approach in an EJ strategy it adopted in 1995. In some areas, such as agricultural chemicals and effects to migrant workers, USDA reviews its practices to identify potential disproportionate, adverse impacts on EJ communities, according to Blake Velde, Senior Environmental Scientist with the USDA Hazardous Materials Management Division. Generally, however, USDA believes its existing technical and financial assistance programs provide solutions to environmental inequity, such as its initiatives on education, food deserts, and economic development in impacted communities, and ensuring access to environmental benefits is the focus of USDA's EJ efforts.

Natural Resources and Environment (NRE) Under Secretary Harris Sherman is the political appointee generally responsible for USDA's EJ strategy, with Patrick Holmes, a senior staffer to the Under Secretary, playing a coordinating role. Although USDA has no staff dedicated solely to EJ, its sub-agencies have many offices dedicated to civil rights compliance, outreach and communication and environmental review whose responsibilities incorporate EJ issues. The Strategic Plan was developed with the input of an Environmental Justice Working Group, made up of staff and leadership representing the USDA's seven mission areas and the SES-level contacts, which were appointed in early 2012, serve as a steering committee for the agency's efforts. The Strategic Plan is organized according to six goals, which were purposefully left broad, and lists specific objectives and agency performance measures under each goal. The details and specific implementation of many of these programs and the performance measures are left to the departments and sub-agencies to develop.

The six goals are to:

- Ensure USDA programs provide opportunities for EJ communities.
- Provide targeted training and capacity-building to EJ communities.
- Expand public participation in agency activities, to enhance the "credibility and public trust" of the USDA.
- Ensure USDA's activities do not have disproportionately high and adverse human health impacts, and resolve environmental justice issues and complaints.
- Increase the awareness of EJ issues among USDA employees.
- Update and/or Develop Departmental and Agency Regulations on EJ.

The Strategic Plan also lists existing programs that either currently support the goal, or are expected to in the future. According to Holmes, some of the challenges of the Strategic Plan process have stemmed from the diverse programs and missions that the agency serves, limitations on staff time, and budgets.

Environmental Justice Initiatives

The Strategic Plan requires that EJ must be integrated into the strategies and evaluations for sub-agencies' technical and financial assistance programs. It also emphasizes public participation, community capacity-building, EJ awareness and training within the USDA.

Transparency, Accountability, Accessibility and Community Participation

A stated goal of USDA's Strategic Plan is to expand public participation in agency activities, to enhance the "credibility and public trust" of the USDA. Specifically, the agency will update its public participation guidelines to include EJ, beginning this process by April 15, 2012. The Strategic Plan emphasizes capacity-building in EJ communities, and includes objectives that emphasize communication between USDA and environmental justice communities, including Tribal consultation. Sub-agencies must announce schedules for training programs in EJ communities and to develop new, preliminary outreach materials on USDA programs by April 15, 2012. An additional performance standard is to encourage EJ communities to participate in the NEPA process, an effort the Strategic Plan requires on or before February 29, 2012, although the Strategic Plan does not articulate a standard by which this could be measured. The Strategic Plan also reiterates compliance with the Executive Orders on Tribal consultation and outreach to non-proficient English speakers, and seeks more diverse representation on regional forest advisory committees [community participation, outreach].

Generally, the USDA's process for developing the Strategic Plan demonstrates a commitment to public involvement. The USDA EJ documents are currently housed obscurely within the Departmental Management section of the USDA website, under the Hazardous Materials Management Division, although the agency plans to update its entire site in 2012 and create a more robust EJ page. The Strategic Plan was released in draft form in December 2011 for a 30-day public comment period, and responses to general types of comments received are in the Progress Report, although the comments themselves are not online. The Secretary's message accompanying the Strategic

Plan requests that organizations and individuals to continue to contact USDA with comments on the Strategic Plan and to identify USDA programs that have been the most beneficial to their communities. The agency has a dedicated email address for this purpose. Agency leadership has asked its sub-agencies to prepare responses to additional comments that have been received, and the agency will release an interim progress report, prior to winter 2013. [community participation, outreach, education].

Internal Evaluation and Training

The Strategic Plan also seeks to increase the awareness of environmental justice issues among USDA employees. The Strategic Plan does not list any existing programs in this area, but does list a series of performance measures going forward, most of which must be met by April 15, 2012. The measures include environmental justice trainings, new web pages, and potential revisions to staff manuals and handbooks. Sub-agencies began reviewing their existing training in 2012 and in their April 9, 2012 reports to the USDA EJ steering committee, sub-agencies were asked to describe their goals for enhanced EJ training. This internal, educational undertaking appears to be new in the 2012 Strategic Plan. The Strategic Plan targets Responsible Officials, meaning office and program managers, for the trainings, as well as the SES-level points of contact required by the Secretary's message. [education, study, compliance and enforcement].

The EJ Strategy tasked each sub-agency with developing its own EJ strategy document by spring 2012, although as of May 2012 the sub-agencies were still in an evaluation stage and had not issued final documents. For many sub-agencies, the 2012 process has been their first focused assessment of their EJ impact and opportunities. Going forward, sub-agencies will submit twice-yearly reports to NRE about their implementation of the Strategic Plan's goals; the first of these was due April 9, 2012, and as of May 2012, the USDA's EJ steering committee was evaluating the first reports.

Establishment of Performance Metrics

As part of its effort to ensure that EJ communities have the opportunity to participate in USDA programs, the Strategic Plan requires each sub-agency to set measurements through which it can track increased EJ community participation in USDA technical and financial assistance programs. This must be done by April 15, 2012. As of late April 2012, the sub-agencies were still in the process of describing a baseline of current activities and determining the metrics to evaluate improvement, such as staff time, grant funding or increased programming. The ultimate metrics are likely to be somewhat subjective, and must be flexible given the broad range of undertakings by the sub-agencies. Also related to evaluation, the Strategic Plan requires the sub-agencies to determine an effective methodology with which they can evaluate whether USDA programs have disproportionate impacts. [study, redressing environmental racism, compliance and enforcement].

Other Ej Initiatives

Tribal Outreach

USDA has had a role in implementing Michelle Obama's *Let's Move* campaign in Tribal Areas, by increasing participation by Bureau of Indian Education schools in Federal nutrition programs, in the development of community gardens on Tribal lands, and in the development of Tribal food policy councils. This is combined with measures to provide Rural Development funding for community infrastructure in Indian Country. [children's issues, education, diet, grants, Native Americans, public health].

The U.S. Forest Service (USFS) is working to update its policy on protection and management of Native American Sacred Sites, an effort that has included listening sessions and government-to-government consultation. The Animal and Plant Health Inspection Service (APHIS) has also consulted with Tribes regarding management of reintroduced of species, where Tribes may have a history of subsistence-level hunting of those species. Meanwhile, the Agricultural Marketing Service (AMS) is exploring a program to use meat from bisons raised on Tribal land to supply AMS food distribution programs to Tribes. [Native Americans, diet, subsistence, community participation].

The Intertribal Technical Assistance Network works to improve access of Tribal governments, communities and individuals to USDA technical assistance programs.

Technical and Financial Assistance to Farmers

The Progress Report highlights the NRCS Strike Force Initiative, which has identified impoverished counties in Mississippi, Georgia and Arkansas to receive increased outreach and training regarding USDA assistance programs. USDA credits this increased outreach with generating a 196 percent increase in contracts, representing more than 250,000 acres of farmland, in its Environmental Quality Incentives Program. [economic benefit, equitable development, grants, outreach, ej as evaluation criteria] NRCS works with "private landowners protect their natural resources" through conservation planning and assistance with the goal of maintaining "productive lands and healthy ecosystems." NRCS has its own civil rights compliance guidance document, and in 2001 NRCS funded and published a study, "Environmental Justice: Perceptions of Issues, Awareness and Assistance," focused on rural, Southern "Black Belt" counties and analyzing how the NRCS workforce could more effectively integrate environmental justice into impacted communities. [compliance and enforcement, redressing environmental racism, grants, study, ej as evaluation criteria].

The Farm Services Agency in 2011 devoted $100,000 of its Socially Disadvantaged Farmers and Ranchers program budget to improving its outreach to counties with persistent poverty, including improving its materials and building relationships with local universities and community groups. [economic benefit, equitable development, grants, outreach, ej as evaluation criteria].

In addition, USDA's Risk Management Agency has initiated education and outreach to low-income farmers regarding use of biological controls, rather than pesticides, for pest control, efforts that the agency believes are valuable in the face of climate change. [climate change, agricultural chemicals, education].

Green Jobs and Capacity Building

A 2011 MOU between a USDA sub-agency, the Food Safety Inspection Service (FSIS) and the American Indian Science and Engineering Society that aims to increase the number of Native Americans entering the FSIS career path; [education, community participation, economic benefit, green jobs, Native Americans, diet, interagency collaboration].

A partnership between APHIS and the Rural Coalition (Coalicion)--an alliance of regionally and culturally diverse organizations working to build a more just and sustainable food system. The partnership focuses on outreach, fair returns to minority and other small farmers and rural communities, farmworker working conditions, environmental protection and food safety. [agricultural chemicals, community participation, diet, economic benefit, outreach, improving health and safety, ej as evaluation criteria].

USFS is also funding pilot initiatives, such as its Urban Water Ambassadors, summer internship positions for youth who coordinate and implement urban tree planting projects. In 2011, USFS provided a grant to the Maryland Department of Natural Resources that funded 14 summer jobs for youth in Baltimore to work on urban watershed restoration programs. [community participation, green jobs, mapping, water].

Mapping

USFS has established several Urban Field Stations, to research urban natural resources' structure, function, stewardship, and benefits. By mapping urban tree coverage, the agency hopes to identify and prioritize EJ communities for urban forest projects. [community education, mapping, diet, improving health and safety, ej as evaluation criteria].

Another initiative highlighted by the agency is the Food and Nutrition Service and Economic Research Service's Food Desert Locator. The Locator provides a spatial view of food deserts, defined as a low-income census tract where a substantial number or share of residents has low access to a supermarket or large grocery store. It also shows, by census tract, the number and percentage of certain populations, such as children, seniors, or households without a vehicle, with low access to grocery stores. The mapped deserts can be used to direct agency resources to increase access to fresh fruits and vegetables and other food assistance programs, according to Blake Velde, an agency scientist and spokesperson on EJ issues. [diet, mapping, improving health and safety, study, ej as evaluation criteria, services and data available to others].

Rural Outreach

USDA Secretary Tom Vilsack has placed a clear emphasis on supporting EJ in rural areas. Although "often the highest profile battles on [environmental justice] issue[s] are waged in at-risk neighborhoods in major cities or at Superfund sites located near populated urban and suburban areas" Vilsack highlighted the often overlooked rural areas where environmental justice is largely ignored.

Through its Rural Utilities Service, the USDA supports a number of Water and Environmental Programs. These programs work to administer water and wastewater loans or grants to rural areas and cities to support water and wastewater, stormwater and solid waste disposal systems, including SEARCH grants that are targeted to financially distressed, small rural communities and other opportunities specifically for Alaskan Native villages and designated Colonias.; In his speech, Secretary Vilsack said that the USDA funded 2,575 clean water projects in rural areas during a two-year period to address problems ranging from wastewater treatment to sewage treatment. [water, land use, compliance and enforcement, improving health and safety, pollution cleanup, ej as evaluation criteria].

The USDA also supports the Rural Energy for America Grant Program. This program provides grants and loans to farmers, ranchers and rural small businesses to finance renewable energy systems and energy efficiency improvements.[grants, economic benefit, ej as evaluation criteria].

Regulations or Formalized EJ Guidelines

In 1997 the USDA promulgated a departmental regulation providing "direction to [sub-]agencies for integrating environmental justice considerations into USDA programs and activities" (DR 5600-002). Issuance of this regulation was a primary goal of USDA's 1995 EJ strategy document. DR 5600-002 includes guidelines for consideration of EJ in the NEPA process, but also stated that "efforts to address environmental justice are not limited to NEPA compliance." It requires evaluation of activities for potential disproportionate EJ impacts, outreach, and performance-metric based evaluation and reporting on sub-agencies' implementation of EJ goals. DR 5600-002 is a forward-looking, permanent directive that applies to all USDA programs and activities. However, it was not published in the Federal Register as a formal rulemaking and does not create a private right of action or enforcement tool. A Strategic Plan goal is to update this regulation, as well as other departmental regulations and policies on EJ. According to USDA, the EJ definition in DR 5600-002 will be modified in 2012—EJ to include measures to avoid disproportionate negative impacts as well as quality-of-life improvements that the agency believes can benefit impacted communities.

The Strategic Plan also has established a performance standard requiring that existing and new USDA regulations are evaluated for EJ impacts or benefits. Sub-agencies are required to develop a process for this evaluation by April 15, 2012. This performance standard reflects a requirement in DR 5600-002 that required the USDA departmental regulation on rulemaking, DR 1521-1, to be revised to require an EJ evaluation in the rulemaking process. As of 2012, DR 1521-1 requires that a cost-benefit analysis of major human health, safety and environmental regulations include analysis of risks to "persons who are disproportionately exposed or particularly sensitive," although DR 1521-1 does not mention EJ or impacts to minority or low-income communities explicitly. [Land Use - permitting, community participation, compliance and enforcement, study].

Enforcement

The Strategic Plan sets an enforcement-specific goal, which includes objectives to "effectively resolve or adjudicate all environmental justice-related Title VI complaints" and to include environmental justice as a key component of civil rights compliance reviews. Agencies are also required to identify an assessment methodology by April 15, 2012, which can be used to determine whether programs have disproportionately high and adverse environmental and human health impacts. The NRCS has published and updated a Civil Rights Compliance Review Guide, which guides the NRCS Civil Rights Division's review of the compliance with Title VI and 12898 in the agency's state offices, field offices and other facilities. The guide was updated in November 2011 and it does not mention EJ explicitly. However, the Strategic Plan identifies the NRCS compliance review and other outreach and research programs as supporting its EJ enforcement goals.

NEPA

The 1997 Regulation, DR 5600-2 required USDA sub-agencies to develop their own NEPA environmental justice guidance documents. The sub-agencies have done so, with some additional details, such as a reminder that the EJ community should be involved in identifying the alternatives, suggested stakeholders and resources, and guidance to hold meetings at times when working people can get to them, and to translate notices. However, when DR 5600-02 is updated as required by the Strategic Plan, changes could be made to the NEPA section of the Regulation. The Strategic Plan sets a performance standard to encourage interested environmental justice communities to be involved in the public participation process for NEPA documents, although the Strategic Plan does not require updates to the NEPA portions of DR 5600-02.

Although the USDA has integrated EJ into each step of the NEPA process as required by Executive Order 12898, many of the NEPA documents completed by the USDA include only cursory analysis of environmental justice effects. This analysis most often includes a rote paragraph as to what Executive Order 12898 requires and a quick conclusion that the agency action does not affect minority and low-income populations. Some examples where the USDA included more in-depth analysis are:

- Descriptions of the minority and low-income populations that live in the study area.
- Impacts relevant to socio-economic environment including changes in employment and income variations in the distribution of social welfare. [community participation, education, outreach, ej as evaluation criteria].

Permitting

The USDA does not have any permitting initiatives specific to EJ.

Title VI

The USDA has an Office of the Assistant Secretary for Civil Rights whose mission it is to provide leadership and direction "for the fair and equitable treatment of all USDA customers."

In 2003 the USDA revised DR 4300-4, internal regulations requiring a Civil Rights Impact Analysis of all "policies, actions or decisions" affecting the USDA's federally conducted and federally assisted programs or activities. The analysis is used to determine the "scope, intensity, direction, duration, and significance of the effects of an agency's proposed . . . policies, actions or decisions." USDA's departmental regulation on EJ, DR 5600-002, required DR 4300-4 to be revised to "require that Civil Rights Impact Analyses include a finding as to whether proposed or new actions have or do not have a disproportionately high and adverse effect on the human health or the environment of minority populations, and whether such effects can be prevented or mitigated." Although DR 4300-4 was revised in 2003, the revised regulation does not explicitly require a finding on adverse environmental or health impacts. [study, compliance and enforcement].

Right-to-Know Movement

A new movement, bent on educating the people, was born after the Bhopal disaster, called the "right-to-know" movement. A series of laws and reports was created, all built to inform the people of the pollutants being dumped into our neighborhoods and atmosphere, and exactly how much of each chemical is being exposed and dumped. The theory behind "right-to-know" is that once people are informed on what is polluting their neighborhood, then they will begin to take action in both bringing down their own emissions, as well as begin to make the companies causing the most pollution, through means such as protests, to take into account their actions.

Emergency Planning and Right to Know Act of 1986

After the Bhopal disaster, where a Union Carbide plant released forty tons of methyl isocyanate into the atmosphere in a village just south of Bhopal, India, the U.S. government passed the Emergency Planning and Right to Know Act of 1986. Introduced by Henry Waxman, the act required all corporations to report their toxic chemical pollution annually, which was then gathered into a report known as the Toxics Release Inventory (TRI). By collecting this data, the government was able to make sure that companies were no longer releasing excessive amounts of deadly toxins into populated areas, so to prevent another incident like that of the thousands of people killed and the tens of thousands of people injured in the Bhopal disaster.

Corporate Toxics Information Report

The Corporate Toxics Information Project (CTIP) was founded on the guidelines that they will "[develop] and [disseminate] information and analysis on corporate releases of pollutants and the consequences for communities." The overarching goal was to help take corporations into account for their pollution habits, by collecting information and putting it in databases so to make it available to the general public. The four goals of the project were to develop 1) corporate rankings, 2) regional reports, based on state, region, and metropolitan areas, 3) industry reports, based on industrial sectors, and 4) to create a web-based resource open to the entire population, that can depict all the collected data. The data collection would be done by the Environmental Protection Agency (EPA) and then analyzed and disseminated by the PERI institute.

One of the biggest projects of CTIP was the Toxic 100. The Toxic 100 is an index of the top 100 air polluters around the United States in terms of the country's largest corporations. The list is based on the EPA's Risk Screening Environmental Indicators (RSEI), which "assesses the chronic human health risk from industrial toxic releases", as well as the Toxics Release Inventory (TRI), which is where the corporations must report their chemical releases to the US government. Since its original publishing date in 2004, the Toxic 100 has been updated four more times, with the latest publishing date being August 2013.

Around the World

In recent years Environmental Justice campaigns have also emerged in other parts of the world, such as India, South Africa, Israel, Nigeria, Mexico, Hungary, Uganda, and the United Kingdom. In Europe for example, there is evidence to suggest that the Romani people and other minority groups of non-European descent are suffering from environmental inequality and discrimination.

In Europe

In Europe, the Romani peoples are ethnic minorities and differ from the rest of the European people by their culture, language, and history. The environmental discrimination that they experience ranges from the unequal distribution of environmental harms as well as the unequal distribution of education, health services and employment. In many countries Romani peoples are forced to live in the slums because many of the laws to get residence permits are discriminatory against them. This forces Romani people to live in urban "ghetto" type housing or in shantytowns. In the Czech Republic and Romania, the Romani peoples are forced to live in places that have less access to running water and sewage, and in Ostrava, Czech Republic, the Romani people live in apartments located above an abandoned mine, which emits methane. Also in Bulgaria, the public infrastructure extends throughout the town of Sofia until it reaches the Romani village where there is very little water access or sewage capacity.

The European Union is trying to strive towards environmental justice by putting into effect declarations that state that all people have a right to a healthy environment. The Stockholm Declaration, the 1987 Brundtland Commission's Report – "Our Common Future", the Rio Declaration, and Article 37 of the Charter of Fundamental Rights of the European Union, all are ways that the Europeans have put acts in place to work toward environmental justice. Europe also funds action-oriented projects that work on furthering Environmental Justice throughout the world. For example, EJOLT (Environmental Justice Organisations, Liabilities and Trade) is a large multinational project supported through the FP7 Science in Society budget line from the European Commission. From March 2011 to March 2015, 23 civil society organizations and universities from 20 countries in Europe, Africa, Latin-America, and Asia are, and have promised to work together on advancing the cause of Environmental Justice. EJOLT is building up case studies, linking organisations worldwide, and making an interactive global map of Environmental Justice.

In the United Kingdom

Whilst the predominant agenda of the Environmental Justice movement in the United States has been tackling issues of race, inequality, and the environment, environmental justice campaigns around the world have developed and shifted in focus. For example, the EJ movement in the United Kingdom is quite different. It focuses on issues of poverty and the environment, but also tackles issues of health inequalities and social exclusion. A UK-based NGO, named the Environmental Justice Foundation, has sought to make a direct link between the need for environmental security and the defense of basic human rights. They have launched several high profile campaigns that link environmental problems and social injustices. A campaign against illegal, unreported and unregulated (IUU) fishing highlighted how 'pirate' fisherman are stealing food from local, artisanal fishing communities. They have also launched a campaign exposing the environmental and human rights abuses involved in cotton production in Uzbekistan. Cotton produced in Uzbekistan is often harvested by children for little or no pay. In addition, the mismanagement of water resources for crop irrigation has led to the near eradication of the Aral Sea. The Environmental Justice Foundation has successfully petitioned large retailers such as Wal-mart and Tesco to stop selling Uzbek cotton.

Building of Alternatives to Climate Change

In France, numerous Alternatiba events, or villages of alternatives, are providing hundreds of alternatives to climate change and lack of environmental justice, both in order to raise people's awareness and to stimulate behaviour change. They have been or will be organized in over sixty different French and European cities, such as Bilbao, Brussels, Geneva, Lyon or Paris.

In South Africa

Under colonial and apartheid governments in South Africa, thousands of black South Africans were removed from their ancestral lands to make way for game parks. Earthlife Africa was formed in 1988 (www.earthlife.org.za), making it Africa's first environmental justice organisation. In 1992, the Environmental Justice Networking Forum (EJNF), a nationwide umbrella organization designed to coordinate the activities of environmental activists and organizations interested in social and environmental justice, was created. By 1995, the network expanded to include 150 member organizations and by 2000, it included over 600 member organizations.

With the election of the African National Congress (ANC) in 1994, the environmental justice movement gained an ally in government. The ANC noted "poverty and environmental degradation have been closely linked" in South Africa. The ANC made it clear that environmental inequalities and injustices would be addressed as part of the party's post-apartheid reconstruction and development mandate. The new South African Constitution, finalized in 1996, includes a Bill of Rights that grants South Africans the right to an "environment that is not harmful to their health or well-being" and "to have the environment protected, for the benefit of present and future generations through reasonable legislative and other measures that

1. prevent pollution and ecological degradation;
2. promote conservation; and
3. secure ecologically sustainable development and use of natural resources while promoting justifiable economic and social development".

South Africa's mining industry is the largest single producer of solid waste, accounting for about two-thirds of the total waste stream. Tens of thousands of deaths have occurred among mine workers as a result of accidents over the last century. There have been several deaths and debilitating diseases from work-related illnesses like asbestosis. For those who live next to a mine, the quality of air and water is poor. Noise, dust, and dangerous equipment and vehicles can be threats to the safety of those who live next to a mine as well. These communities are often poor and black and have little choice over the placement of a mine near their homes. The National Party introduced a new Minerals Act that began to address environmental considerations by recognizing the health and safety concerns of workers and the need for land rehabilitation during and after mining operations. In 1993, the Act was amended to require each new mine to have an Environmental Management Program Report (EMPR) prepared before breaking ground. These EMPRs were intended to force mining companies to outline all the possible environmental impacts of the particular mining operation and to make provision for environmental management.

In October 1998, the Department of Minerals and Energy released a White Paper entitled *A Minerals and Mining Policy for South Africa*, which included a section on Environmental Management. The White Paper states "Government, in recognition of the responsibility of the State as custodian of the nation's natural resources, will ensure that the essential development of the country's mineral resources will take place within a framework of sustainable development and in accordance with national environmental policy, norms, and standards". It adds that any environmental policy "must ensure a cost-effective and competitive mining industry."

In Australia

In Australia, the "Environmental Justice Movement" is not defined as it is in the United States. Australia does have some discrimination mainly in the siting of hazardous waste facilities in areas where the people are not given proper information about the company. The injustice that takes place in Australia is defined as environmental politics on who get the unwanted waste site or who has control over where factory opens up. The movement towards equal environmental politics focuses more on who can fight for companies to build, and takes place in the parliament; whereas, in the United States Environmental Justice is trying to make nature safer for all people.

In Ecuador

An example of the environmental injustices that indigenous groups face can be seen in the Chevron-Texaco incident in the Amazon rainforest. Texaco, which is now Chevron, found oil in Ecuador in 1964 and built sub-standard oil wells to cut costs. The deliberately used inferior technology to make their operations cheaper, even if detrimental to the local people and environment. After the company left in 1992, they left approximately one thousand toxic waste pits open and dumped billions of gallons of toxic water into the rivers.

In South Korea

South Korea has a relatively short history of environmental justice compared to other countries in the west. As a result of rapid industrialization, people started to have awareness on pollution, and from the environmental discourses the idea of environmental justice appeared. The concept of environmental justice appeared in South Korea in late 1980s.

South Korea experienced rapid economic growth (which is commonly referred to as the 'Miracle on the Han River') in the 20th century as a result of industrialization policies adapted by Park Chung-hee after 1970s. The policies and social environment had no room for environmental discussions, which aggravated the pollution in the country.

Environmental movements in South Korea started from air pollution campaigns. As the notion of environment pollution spread, the focus on environmental activism shifted from existing pollution to preventing future pollution, and the organizations eventually started to criticize the government policies that are neglecting the environmental issues. The concept of environmental justice was introduced in South Korea among the discussions of environment after 1990s. While the environmental organizations analyzed the condition of pollution in South Korea, they noticed that the environmental problems were inequitably focused especially on regions where people with low social and economic status were concentrated.

The problems of environmental injustice have arisen by environment related organizations, but approaches to solve the problems were greatly supported by the government, which developed various policies and launched institution. These actions helped raise awareness of environmental justice in South Korea. Existing environment policies were modified to cover environmental justice issues.

Environmental justice began to be widely recognized in the 1990s through policy making and researches of related institutions. For example, the Ministry of Environment, which was founded in 1992, launched Citizen's Movement for Environmental Justice (CMEJ) to raise awareness of the problem and figure out appropriate plans. As a part of its activities, Citizen's Movement for Environmental Justice (CMEJ) held Environmental Justice forum in 1999, to gather and analyze the existing studies on the issue which were done sporadically by various organizations. Citizen's Movement for Environmental Justice (CMEJ) started as a small organization, but it is keep growing and expanding. In 2002, CMEJ had more than 5 times the numbers of members and 3 times the budget it had in the beginning year.

Environmental injustice is still an ongoing problem. One example is the construction of Saemangeum Seawall. The construction of Saemangeum Seawall, which is the world's longest dyke (33 kilometers) runs between Yellow Sea and Saemangeum estuary, was part of a government project initiated in 1991. The project raised concerns on the destruction of ecosystem and taking away the local residential regions. It caught the attention of environmental justice activists because the main victims were low-income fishing population and their future generations. This is considered as an example of environmental injustice which was caused by the execution of exclusive development-centered policy.

The construction of Seoul-Incheon canal also raised environmental justice controversies. The construction took away the residential regions and farming areas of the local residents. Also, the environment worsened in the area because of the appearance of wet fogs which was caused by water deprivation and local climate changes caused by the construction of canal. The local residents, mostly people with weak economic basis, were severely affected by the construction and became the main victims of such environmental damages. While the socially and economically weak citizens suffered from the environmental changes, most of the benefits went to the industries and conglomerates with political power.

Construction of industrial complex was also criticized in the context of environmental justice. The conflict in wicheon region is one example. The region became the center of controversy when the government decided to build industrial complex of dye houses, which were formerly located in Daegu metropolitan region. As a result of the construction, Nakdong River, which is one of the main rivers in South Korea, were contaminated and local residents suffered from environmental changes caused by the construction.

Environmental justice is a growing issue in South Korea. Although the issue is not yet widely recognized compared to other countries, many organizations beginning to recognize the issue.

Between Northern and Southern Countries

Environmental discrimination in a global perspective is also an important factor when examining the Environmental Justice movement. Even though the Environmental Justice movement began in the United States, the United States also contributes to expanding the amount of environmental injustice that takes place in less-developed countries. Some companies in the United States and in other developed nations around the world contribute to the injustice by shipping the toxic waste and byproducts of factories to less-developed countries for disposal. This act increases the amount of waste in the third world countries, most of which do not have proper sanitation for their own waste much less the waste of another country. Often, the people of the less-developed countries are exposed to toxins from this waste and do not even realize what kind of waste they are encountering or the health problems that could come with it.

One prominent example of northern countries shipping their waste to southern countries took place in Haiti. Philadelphia, Pennsylvania had ash from the incineration of toxic waste that they did not have room to dump. Philadelphia decided to put the ash into the hands of a private company, which shipped the ash and dumped it in various other parts of the world, outside of the United States. *The Khian Sea*, the ship the ash was put on, sailed around the world and many countries would not accept the waste because it was hazardous for the environment and the people. The ship owners finally dumped the waste, labeled Fertilizer, in Haiti, on the beach, and sailed away in the night. The government of Haiti was infuriated and called for the waste to be removed, but the company would not come to take the ash away. The fighting over who was responsible for the waste and who would remove the waste went on for many years. After debating for over ten years, the waste was removed and taken back to a site just outside Philadelphia to be disposed of permanently.

The reason that this transporting of waste from Northern countries to the Southern countries takes place is because it is cheaper to transport waste to another country and dump it there, than to pay to dump the waste in the producing country because the third world countries do not have the same strict industry regulations as the more developed countries. The countries that the waste is taken to are usually empoverished and the governments have little or no control over the happenings in the country or do not care about the people.

Transnational Movement Networks

Many of the Environmental Justice Networks that began in the United States expanded their horizons to include many other countries and became Transnational Networks for Environmental Justice. These networks work to bring Environmental Justice to all parts of the world and protect all citizens of the world to reduce the environmental injustice happening all over the world. Listed below are some of the major Transnational Social Movement Organizations.

- Basel Action Network — works to end toxic waste dumping in poor undeveloped countries from the rich developed countries.
- GAIA (Global Anti-Incinerator Alliance) — works to find different ways to dispose of waste other than incineration. This company has people working in over 77 countries throughout the world.
- GR (Global Response) — works to educate activists and the upper working class how to protect human rights and the ecosystem.
- Greenpeace International — which was the first organization to become the global name of Environmental Justice. Greenpeace works to raise the global consciousness of transnational trade of toxic waste.
- Health Care without Harm — works to improve the public health by reducing the environmental impacts of the health care industry.
- International Campaign for Responsible Technology — works to promote corporate and government accountability with electronics and how the disposal of technology affect the environment.
- International POPs Elimination Network — works to reduce and eventually end the use of persistent organic pollutants (POPs) which are harmful to the environment.

- PAN (Pesticide Action Network) — works to replace the use of hazardous pesticides with alternatives that are safe for the environment.

These global networks work together to achieve the shared goal of a cleaner environment.

Environmental Governance

Environmental governance is a concept in political ecology and environmental policy that advocates sustainability (sustainable development) as the supreme consideration for managing all human activities—political, social and economic. Governance includes government, business and civil society, and emphasizes whole system management. To capture this diverse range of elements, environmental governance often employs alternative systems of governance, for example watershed-based management.

It views natural resources and the environment as global public goods, belonging to the category of goods that are not diminished when they are shared. This means that everyone benefits from for example, a breathable atmosphere, stable climate and stable biodiversity.

Public goods are non-rivalrous—a natural resource enjoyed by one person can still be enjoyed by others—and non-excludable—it is impossible to prevent someone consuming the good (breathing). Nevertheless, public goods are recognized as beneficial and therefore have value. The notion of a global public good thus emerges, with a slight distinction: it covers necessities that must not be destroyed by one person or state.

The non-rivalrous character of such goods calls for a management approach that restricts public and private actors from damaging them. One approach is to attribute an economic value to the resource. Water is possibly the best example of this type of good.

As of 2013 environmental governance is far from meeting these imperatives. "Despite a great awareness of environmental questions from developed and developing countries, there is environmental degradation and the appearance of new environmental problems. This situation is caused by the parlous state of global environmental governance, wherein current global environmental governance is unable to address environmental issues due to many factors. These include fragmented governance within the United Nations, lack of involvement from financial institutions, proliferation of environmental agreements often in conflict with trade measures; all these various problems disturb the proper functioning of global environmental governance. Moreover, divisions among northern countries and the persistent gap between developed and developing countries also have to be taken into account to comprehend the institutional failures of the current global environmental governance."

Definitions

What is Environmental Governance?

Environmental governance refers to the processes of decision-making involved in the control and management of the environment and natural resources. International Union for Conservation of

Nature (IUCN), define Environmental Governance as the 'Multi-level interactions (i.e., local, national, international/global) among, but not limited to, three main actors, i.e., state, market, and civil society, which interact with one another, whether in formal and informal ways; in formulating and implementing policies in response to environment-related demands and inputs from the society; bound by rules, procedures, processes, and widely accepted behavior; possessing characteristics of "good governance"; for the purpose of attaining environmentally-sustainable development' (ICUN 2014).

Key principles of environmental governance include:

- Embedding the environment in all levels of decision-making and action.
- Conceptualizing cities and communities, economic and political life as a subset of the environment.
- Emphasizing the connection of people to the ecosystems in which they live.
- Promoting the transition from open-loop/cradle-to-grave systems (like garbage disposal with no recycling) to closed-loop/cradle-to-cradle systems (like permaculture and zero waste strategies).

Neoliberal Environmental Governance – is an approach to the theory of environmental governance framed by a perspective on neoliberalism as an ideology, policy and practice in relation to the biophysical world. There are many definitions and applications of neoliberalism, e.g. in economic, international relations, etc. However, the traditional understanding of neoliberalism is often simplified to the notion of the primacy of market-led economics through the rolling back of the state, deregulation and privatisation. Neoliberalism has evolved particularly over the last 40 years with many scholars leaving their ideological footprint on the neoliberal map. Hayek and Friedman believed in the superiority of the free market over state intervention. As long as the market was allowed to act freely, the supply/demand law would ensure the 'optimal' price and reward. In Karl Polanyi's opposing view this would also create a state of tension in which self-regulating free markets disrupt and alter social interactions and "displace other valued means of living and working". However, in contrast to the notion of an unregulated market economy there has also been a "paradoxical increase in [state] intervention" in the choice of economic, legislative and social policy reforms, which are pursued by the state to preserve the neoliberal order. This contradictory process is described by Peck and Tickell as roll back/roll out neoliberalism in which on one hand the state willingly gives up the control over resources and responsibility for social provision while on the other, it engages in "purposeful construction and consolidation of neoliberalised state forms, modes of governance, and regulatory relations".

There has been a growing interest in the effects of neoliberalism on the politics of the non-human world of environmental governance. Neoliberalism is seen to be more than a homogenous and monolithic 'thing' with a clear end point. It is a series of path-dependent, spatially and temporally "connected neoliberalisation" processes which affect and are affected by nature and environment that "cover a remarkable array of places, regions and countries". Co-opting neoliberal ideas of the importance of private property and the protection of individual (investor) rights, into environmental governance can be seen in the example of recent multilateral trade agreements (see in particular the North American Free Trade Agreement). Such neoliberal structures further reinforce a process of nature enclosure and primitive accumulation or "accumulation by dispossession" that serves to privatise increasing areas of nature. The ownership-transfer of resources traditionally not privately owned to free market mechanisms is believed to deliver greater efficiency and optimal return on investment. Other similar examples of neo-liberal inspired projects include the enclosure of minerals, the fisheries quota system in the North Pacific and the privatisation of water supply and sewage treatment in England and Wales. All three examples share neoliberal characteristics to "deploy markets as the solution to environmental problems" in which scarce natural resources are commercialized and turned into commodities. The approach to frame the ecosystem in the context of a price-able commodity is also present in the work of neoliberal geographers who subject nature to price and supply/demand mechanisms where the earth is considered to be a quantifiable resource (Costanza, for example, estimates the earth ecosystem's service value to be between 16 and 54 trillion dollars per year).

Environmental Issues

Main Drivers of Environmental Degradation

Economic growth – The development-centric vision that prevails in most countries and international institutions advocates a headlong rush towards more economic growth. Environmental economists on the other hand, point to a close correlation between economic growth and environmental degradation, arguing for qualitative development as an alternative to growth. As a result, the past couple of decades has seen a big shift towards sustainable development as an alternative to neo-liberal economics. There are those, particularly within the alternative globalization movement, who maintain that it is feasible to change to a degrowth phase without losing social efficiency or lowering the quality of life.

Consumption – The growth of consumption and the cult of consumption, or consumerist ideology, is the major cause of economic growth. Overdevelopment, seen as the only alternative to poverty, has become an end in itself. The means for curbing this growth are not equal to the task, since the phenomenon is not confined to a growing middle class in developing countries, but also concerns the development of irresponsible lifestyles, particularly in northern countries, such as the increase in the size and number of homes and cars per person.

Destruction of biodiversity – The complexity of the planet's ecosystems means that the loss of any species has unexpected consequences. The stronger the impact on biodiversity, the stronger the likelihood of a chain reaction with unpredictable negative effects. Another important factor of environmental degradation that falls under this destruction of biodiversity, and must not be ignored is deforestation. Despite all the damage inflicted, a number of ecosystems have proved to be resilient. Environmentalists are endorsing a precautionary principle whereby all potentially damaging activities would have to be analyzed for their environmental impact.

Population growth – Forecasts predict 8.9 billion people on the planet in 2050. This is a subject which primarily affects developing countries, but also concerns northern countries; although their demographic growth is lower, the environmental impact per person is far higher in these countries. Demographic growth needs to be countered by developing education and family planning programs and generally improving women's status.

"Pollution" - Pollution caused by the use of fossil fuels is another driver of environmental destruction. The burning of carbon-based fossil fuels such as coal and oil, releases carbon dioxide into the atmosphere. One of the major impacts of this is the climate change that is currently taking place on the planet, where the earth's temperature is gradually rising. Given that fuels such as coal and oil are the most heavily used fuels, this a great concern to many environmentalists.

"Agricultural practices" - Destructive agricultural practices such as overuse of fertilizers and overgrazing lead to land degradation. The soil gets eroded, and leads to silting in rivers and reservoirs. Soil erosion is a continuous cycle and ultimately results in desertification of the land. Apart from land degradation, water pollution is also a possibility; chemicals used in farming can run-off into rivers and contaminate the water.

Challenges

The crisis by the impact of human activities on nature calls for governance. Which includes responses by international institutions, governments and citizens, who should meet this crisis by pooling the experience and knowledge of each of the agents and institutions concerned.

The environmental protection measures taken remain insufficient. The necessary reforms require time, energy, money and diplomatic negotiations. The situation has not generated a unanimous response. Persistent divisions slow progress towards global environmental governance.

The global nature of the crisis limits the effects of national or sectoral measures. Cooperation is necessary between actors and institutions in international trade, sustainable development and peace.

Global, continental, national and local governments have employed a variety of approaches to environmental governance. Substantial positive and negative spillovers limit the ability of any single jurisdiction to resolve issues.

Challenges facing environmental governance include:

- Inadequate continental and global agreements.
- Unresolved tensions between maximum development, sustainable development and maximum protection, limiting funding, damaging links with the economy and limiting application of Multilateral Environment Agreements (MEAs).
- Environmental funding is not self-sustaining, diverting resources from problem-solving into funding battles.
- Lack of integration of sector policies.
- Inadequate institutional capacities.
- Ill-defined priorities.
- Unclear objectives.
- Lack of coordination within the UN, governments, the private sector and civil society.
- Lack of shared vision.

- Interdependencies among development/sustainable economic growth, trade, agriculture, health, peace and security.
- International imbalance between environmental governance and trade and finance programs, e.g., World Trade Organization (WTO).
- Limited credit for organizations running projects within the Global Environment Facility (GEF).
- Linking UNEP, United Nations Development Programme (UNDP) and the World Bank with MEAs.
- Lack of government capacity to satisfy MEA obligations.
- Absence of the gender perspective and equity in environmental governance.
- Inability to influence public opinion.
- Time lag between human action and environmental effect, sometimes as long as a generation.
- Environmental problems being embedded in very complex systems, of which our understanding is still quite weak.

All of these challenges have implications on governance, however international environmental governance is necessary. The IDDRI claims that rejection of multilateralism in the name of efficiency and protection of national interests conflicts with the promotion of international law and the concept of global public goods. Others cite the complex nature of environmental problems.

tBäckstrand and Saward wrote, "sustainability and environmental protection is an arena in which innovative experiments with new hybrid, plurilateral forms of governance, along with the incorporation of a transnational civil society spanning the public-private divide, are taking place."

Local Governance

A 1997 report observed a global consensus that sustainable development implementation should be based on local level solutions and initiatives designed with and by the local communities. Community participation and partnership along with the decentralisation of government power to local communities are important aspects of environmental governance at the local level. Initiatives such as these are integral divergence from earlier environmental governance approaches which was "driven by state agendas and resource control" and followed a top-down or trickle down approach rather than the bottom up approach that local level governance encompasses. Local level governance shifts decision making power away from the state and/or governments to the grassroots. Local level governance is extremely important even on a global scale. Environmental governance at the global level is defined as international and as such has resulted in the marginalisation of local voices. Local level governance is important to bring back power to local communities in the global fight against environmental degridation. Pulgar Vidal observed a "new institutional framework, [wherein] decision-making regarding access to and use of natural resources has become increasingly decentralized." He noted four techniques that can be used to develop these processes.

- Formal and informal regulations, procedures and processes, such as consultations and participative democracy;
- Social interaction that can arise from participation in development programs or from the reaction to perceived injustice;
- Regulating social behaviours to reclassify an individual question as a public matter;
- Within-group participation in decision-making and relations with external actors.

He found that the key conditions for developing decentralized environmental governance are:

- Access to social capital, including local knowledge, leaders and local shared vision;
- Democratic access to information and decision-making;
- Local government activity in environmental governance: as facilitator of access to natural resources, or as policy maker;
- An institutional framework that favours decentralized environmental governance and creates forums for social interaction and making widely accepted agreements acceptable.

The legitimacy of decisions depends on the local population's participation rate and on how well participants represent that population. With regard to public authorities, questions linked to biodiversity can be faced by adopting appropriate policies and strategies, through exchange of knowledge and experience, the forming of partnerships, correct management of land use, monitoring of biodiversity and optimal use of resources, or reducing consumption, and promoting environmental certifications, such as EMAS and/or ISO 14001. Local authorities undoubtedly have a central role to play in the protection of biodiversity and this strategy is successful above all when the authorities show strength by involving stakeholders in a credible environmental improvement project and activating a transparent and effective communication policy (Ioppolo et al., 2013).

State Governance

States play a crucial role in environmental governance, because "however far and fast international economic integration proceeds, political authority remains vested in national governments". It is for this reason that governments should respect and support the commitment to implementation of international agreements.

At the state level, environmental management has been found to be conducive to the creation of roundtables and committees. In France, the *Grenelle de l'environnement* process:

- Included a variety of actors (e.g. the state, political leaders, unions, businesses, not-for-profit organizations and environmental protection foundations);
- Allowed stakeholders to interact with the legislative and executive powers in office as indispensable advisors;
- Worked to integrate other institutions, particularly the Economic and Social Council, to form a pressure group that participated in the process for creating an environmental governance model;
- Attempted to link with environmental management at regional and local levels.

If environmental issues are excluded from e.g., the economic agenda, this may delegitimize those institutions.

"In southern countries, the main obstacle to the integration of intermediate levels in the process of territorial environmental governance development is often the dominance of developmentalist inertia in states' political mindset. The question of the environment has not been effectively integrated in national development planning and programs. Instead, the most common idea is that environmental protection curbs economic and social development, an idea encouraged by the frenzy for exporting raw materials extracted using destructive methods that consume resources and fail to generate any added value." Of course they are justified in this thinking, as their main concerns are social injustices such as poverty alleviation. Citizens in some of these states have responded by developing empowerment strategies to ease poverty through sustainable development. In addition to this, policymakers must be more aware of these concerns of the global south, and must make sure to integrate a strong focus on social justice in their policies.

Global Governance

At the global level there are numerous important actors involved in environmental governance and "a range of institutions contribute to and help define the practice of global environmental governance. The idea of global environmental governance is to govern the environment at a global level through a range of nation states and non state actors such as national governments, NGOs and other international organisations such as UNEP (United Nations Environment Programme). Global environmental governance is the answer to calls for new forms of governance because of the increasing complexity of the international agenda. It is perceived to be an effective form of multilateral management and essential to the international community in meeting goals of mitigation and the possible reversal of the impacts on the global environment. However, a precise definition of global environmental governance is still vague and there are many issues surrounding global governance. Elliot argues that "the congested institutional terrain still provides more of an appearance than a reality of comprehensive global governance." This meant that there are too many institutions within the global governance of the environment for it to be completely inclusive and coherent leaving it merely portraying the image of this to the global public. Global environmental governance is about more than simply expanding networks of institutions and decision makers. "It is a political practice which simultaneously reflects, constitutes and masks global relations of power and powerlessness." State agendas exploit the use of global environmental governance to enhance their oven agendas or wishes even if this is at the detriment of the vital element behind global environmental governance which is the environment. Elliot states that global environmental governance "is neither normatively neutral nor materially benign." As explored by Newell, report notes by The Global Environmental Outlook noted that the systems of global environmental governance are becoming increasingly irrelevant or impotent due to patterns of globalisation such as; imbalances in productivity and the distribution of goods and services, unsustainable progression of extremes of wealth and poverty and population and economic growth overtaking environmental gains. Newell states that, despite such acknowledgements, the "managing of global environmental change within International Relations continues to look to international regimes for the answers."

Issues of Scale

Multi-Tier Governance

The literature on governance scale shows how changes in the understanding of environmental issues have led to the movement from a local view to recognising their larger and more complicated scale. This move brought an increase in the diversity, specificity and complexity of initiatives. Meadowcroft pointed out innovations that were layered on top of existing structures and processes, instead of replacing them.

Lafferty and Meadowcroft give three examples of multi-tiered governance: internationalisation, increasingly comprehensive approaches, and involvement of multiple governmental entities. Lafferty and Meadowcroft described the resulting multi-tiered system as addressing issues on both smaller and wider scales.

Institutional Fit

Hans Bruyninckx claimed that a mismatch between the scale of the environmental problem and the level of the policy intervention was problematic. Young claimed that such mismatches reduced the effectiveness of interventions. Most of the literature addresses the level of governance rather than ecological scale.

Elinor Ostrom, amongst others, claimed that the mismatch is often the cause of unsustainable management practices and that simple solutions to the mismatch have not been identified.

Considerable debate has addressed the question of which level(s) should take responsibility for fresh water management. Development workers tend to address the problem at the local level. National governments focus on policy issues. This can create conflicts among states because rivers cross borders, leading to efforts to evolve governance of river basins.

Environmental Governance Issues

Soil Deterioration

Soil and land deterioration reduces its capacity for capturing, storing and recycling water, energy and food. Alliance 21 proposed solutions in the following domains:

- Include soil rehabilitation as part of conventional and popular education.
- Involve all stakeholders, including policymakers and authorities, producers and land users, the scientific community and civil society to manage incentives and enforce regulations and laws.
- Establish a set of binding rules, such as an international convention.
- Set up mechanisms and incentives to facilitate transformations.
- Gather and share knowledge.
- Mobilize funds nationally and internationally.

Climate Change

The scientific consensus on climate change is expressed in the reports of Intergovernmental Panel on Climate Change (IPCC) and also in the statements by all major scientific bodies in the United States such as National Academy of Sciences.

The drivers of climate change can include - Changes in solar irradiance - Changes in atmospheric trace gas and aerosol concentrations Evidence of climate change can be identified by examining - Atmospheric concentrations of Green House Gases (GHGs) such as carbon dioxide (CO2) - Land and sea surface temperatures - Atmospheric water vapor - Precipitation - The occurrence or strength of extreme weather and climate events - Glaciers - Rapid sea ice loss - Sea level.

It is suggested by climate models that the changes in temperature and sea level can be the causal effects of human activities such as consumption of fossil fuels, deforestation, increased agricultural production and production of xenobiotic gases.

There has been increasing actions in order to mitigate climate change and reduce its impact at national, regional and international levels. Kyoto protocol and United Nations Framework Convention on Climate Change (UNFCCC) plays the most important role in addressing climate change at an international level.

The goal of combating climate change led to the adoption of the Kyoto Protocol by 191 states, an agreement encouraging the reduction of greenhouse gases, mainly CO_2. Since developed economies produce more emissions per capita, limiting emissions in all countries inhibits opportunities for emerging economies, the only major success in efforts to produce a global response to the phenomenon.

Two decades following the Brundtland Report, however, there has been no improvement in the key indicators highlighted.

Biodiversity

Environmental governance for protecting the biodiversity has to act in many levels. Biodiversity is fragile because it is threatened by almost all human actions. To promote conservation of biodiversity, agreements and laws have to be created to regulate agricultural activities, urban growth, industrialization of countries, use of natural resources, control of invasive species, the correct use of water and protection of air quality. Before making any decision for a region or country decision makers, politicians and community have to take into account what are the potential impacts for biodiversity, that any project can have.

Population growth and urbanization have been a great contributor for deforestation. Also, population growth requires more intense agricultural areas use, which also results in necessity of new areas to be deforested. This causes habitat loss, which is one of the major threats for biodiversity. Habitat loss and habitat fragmentation affects all species, because they all rely on limited resources, to feed on and to breed.

'Species are genetically unique and irreplaceable their loss is irreversible. Ecosystems vary across a vast range of parameters, and similar ecosystems (whether wetlands, forests, coastal

reserves etc) cannot be presumed to be interchangeable, such that the loss of one can be compensated by protection or restoration of another'.

To avoid habitat loss, and consequently biodiversity loss, politicians and lawmakers should be aware of the precautionary principle, which means that before approving a project or law all the pros and cons should be carefully analysed. Sometimes the impacts are not explicit, or not even proved to exist. However, if there is any chance of an irreversible impact happen, it should be taken into consideration.

To promote environmental governance for biodiversity protection there has to be a clear articulation between values and interests while negotiating environmental management plans. International agreements are good way to have it done right.

The Convention on Biological Diversity (CBD) was signed in Rio de Janeiro in 1992 human activities. The CBD's objectives are: "to conserve biological diversity, to use biological diversity in a sustainable fashion, to share the benefits of biological diversity fairly and equitably." The Convention is the first global agreement to address all aspects of biodiversity: genetic resources, species and ecosystems. It recognizes, for the first time, that the conservation of biological diversity is "a common concern for all humanity". The Convention encourages joint efforts on measures for scientific and technological cooperation, access to genetic resources and the transfer of clean environmental technologies.

The Convention on Biological Diversity most important edition happened in 2010 when the Strategic Plan for Biodiversity 2011-2020 and the Aichi Targets, were launched. These two projects together make the United Nations decade on Biodiversity. It was held in Japan and has the targets of '*halting and eventually reversing the loss of biodiversity of the planet*'. The Strategic Plan for Biodiversity has the goal to '*promote its overall vision of living in harmony with nature*' As result '*mainstream biodiversity at different levels. Throughout the United Nations Decade on Biodiversity, governments are encouraged to develop, implement and communicate the results of national strategies for implementation of the Strategic Plan for Biodiversity*'. According to the CBD the five Aichi targets are:

- '*Address the underlying causes of biodiversity loss by mainstreaming biodiversity across government and society*;
- *Reduce the direct pressures on biodiversity and promote sustainable use*;
- *Improve the status of biodiversity by safeguarding ecosystems, species and genetic diversity*;
- *Enhance the benefits to all from biodiversity and ecosystem services*;
- *Enhance implementation through participatory planning, knowledge management and capacity building.*'

Water

The 2003 UN World Water Development Report claimed that the amount of water available over the next twenty years would drop by 30%.

At that time, 40% of the planet's inhabitants did not have access to the minimum necessary for basic hygiene. Over 2.2 million people died in 2000 from diseases linked to contaminated water, or from drowning. In 2004, the UK's WaterAid charity reported that one child died every 15 seconds from water-linked diseases.

According to Alliance 21 "All levels of water supply management are necessary and independent. The integrated approach to the catchment areas must take into account the needs of irrigation and those of towns, jointly and not separately as is often seen to be the case....The governance of a water supply must be guided by the principles of sustainable development."

Australian water resources have always been variable but they are becoming increasingly so with changing climate conditions. Because of how limited water resources are in Australia, there needs to be an effective implementation of environmental governance conducted within the country. Water restrictions are an important policy device used in Australian environmental governance to limit the amount of water used in urban and agricultural environments (Beeton et al. 2006). There is increased pressure on surface water resources in Australia because of the uncontrolled growth in groundwater use and the constant threat of drought. These increased pressures not only affect the quantity and quality of the waterways but they also negatively affect biodiversity. The government needs to create policies that preserve, protect and monitor Australia's inland water. The most significant environmental governance policy imposed by the Australian government is environmental flow allocations that allocate water to the natural environment. The proper implementation of water trading systems could help to conserve water resources in Australia. Over the years there has been an increase in demand for water, making Australia the third largest per capita user of water in the world (Beeton et al. 2006). If this trend continues, the gap between supply and demand will need to be addressed. The government needs to implement more efficient water allocations and raise water rates (UNEP, 2014). By changing public perception to promote the action of reusing and recycling water some of the stress of water shortages can be alleviated. More extensive solutions like desalination plants, building more dams and using aquifer storage are all options that could be taken to conserve water levels but all these methods are controversial. With caps on surface water use, both urban and rural consumers are turning to groundwater use; this has caused groundwater levels to decline significantly. Groundwater use is very hard to monitor and regulate. There is not enough research currently being conducted to accurately determine sustainable yields. Some regions are seeing improvement in groundwater levels by applying caps on bores and the amount of water that consumers are allowed to extract. There have been projects in environmental governance aimed at restoring vegetation in the riparian zone. Restoring riparian vegetation helps increase biodiversity, reduce salinity, prevent soil erosion and prevent riverbank collapse. Many rivers and waterways are controlled by weirs and locks that control the flow of rivers and also prevent the movement of fish. The government has funded fish-ways on some weirs and locks to allow for native fish to move upstream. Wetlands have significantly suffered under restricted water resources with water bird numbers dropping and a decrease in species diversity. The allocation of water for bird breeding through environmental flows in Macquarie Marshes has led to an increase in breeding (Beeton et al. 2006). Because of dry land salinity throughout Australia there has been an increase in the levels of salt in Australian waterways. There has been funding in salt interception schemes which help to improve in-stream salinity levels but whether river salinity has improved or not is still unclear because there is not enough data available yet. High salinity levels are dangerous because they can negatively affect larval and juvenile stages of certain fish.

The introduction of invasive species into waterways has negatively affected native aquatic species because invasive species compete with native species and alter natural habitats. There has been research in producing daughterless carp to help eradicate carp. Government funding has also gone into building in-stream barriers that trap the carp and prevent them from moving into floodplains and wetlands. Investment in national and regional programmes like the Living Murray (MDBC), Healthy Waterways Partnership and the Clean Up the Swan Programme are leading to important environmental governance. The Healthy Rivers programme promotes restoration and recovery of environmental flows, riparian re-vegetation and aquatic pest control. The Living Murray programme has been crucial for the allocation of water to the environment by creating an agreement to recover 500 billion litres of water to the Murray River environment. Environmental governance and water resource management in Australia must be constantly monitored and adapted to suit the changing environmental conditions within the country (Beeton et al. 2006). If environmental programmes are governed with transparency there can be a reduction in policy fragmentation and an increase in policy efficiency (Mclntyre, 2010).

Ozone Layer

On 16 September 1987 the United Nations General Assembly signed the Montreal Protocol to address the declining ozone layer. Since that time, the use of chlorofluorocarbons (industrial refrigerants and aerosols) and farming fungicides such as methyl bromide has mostly been eliminated, although other damaging gases are still in use.

Nuclear Risk

The Nuclear non-proliferation treaty is the primary multilateral agreement governing nuclear activity.

Transgenic Organisms

Genetically modified organisms are not the subject of any major multilateral agreements. They are the subject of various restrictions at other levels of governance. GMOs are in widespread use in the US, but are heavily restricted in many other jurisdictions.

Controversies have ensued over golden rice, genetically modified salmon, genetically modified seeds, disclosure and other topics.

Precautionary Principle

The precautionary principle or precautionary approach states that if an action or policy has a suspected risk of causing harm to the public or to the environment, in the absence of scientific consensus that the action or policy is harmful, the burden of proof that it is not harmful falls on those taking an action. As of 2013 it was not the basis of major multilateral agreements. The Precautionary Principle is put into effect if there is a chance that proposed action may cause harm to the society or the environment. Therefore, those involved in the proposed action must provide evidence that it will not be harmful, even if scientists do not believe that it will cause harm. It falls upon the policymakers to make the optimal decision, if there is any risk, even without any credible scientific evidence. However, taking precautionary action also means that there is an element of cost involved, either social or economic. So if the cost was seen as insignificant the action would be taken without the implementation of the precautionary principle. But often the cost is ignored, which can lead to harmful repercussions. This is often the case with industry and scientists who are primarily concerned with protecting their own interests.

Agreements

Conventions

The main multilateral conventions, also known as Rio Conventions, are as follows:

Convention on Biological Diversity (CBD) (1992–1993): aims to conserve biodiversity. Related agreements include the Cartagena Protocol on biosafety.

United Nations Framework Convention on Climate Change (UNFCC) (1992–1994): aims to stabilize concentrations of greenhouse gases at a level that would stabilize the climate system without threatening food production, and enabling the pursuit of sustainable economic development; it incorporates the Kyoto Protocol.

United Nations Convention to Combat Desertification (UNCCD) (1994–1996): aims to combat desertification and mitigate the effects of drought and desertification, particularly in Africa.

Further Conventions

- Ramsar Convention on Wetlands of International Importance (1971–1975).
- UNESCO World Heritage Convention (1972–1975).
- Convention on International Trade in Endangered Species of Wild Flora and Fauna (CITES) (1973–1975).
- Bonn Convention on the Conservation of Migratory Species (1979–1983).
- Convention on the Protection and Use of Transboundary Watercourses and International Lakes (Water Convention) (1992–1996).
- Basel Convention on the Control of Transboundary Movements of Hazardous Wastes and their Disposal (1989–1992).
- Rotterdam Convention on the Prior Informed Consent Procedures for Certain Hazardous Chemicals and Pesticides in International Trade.
- Stockholm Convention on Persistent Organic Pollutants (COP) (2001–2004).

The Rio Conventions are Characterized by:

- Obligatory execution by signatory states.
- Involvement in a sector of global environmental governance.

- Focus on the fighting poverty and the development of sustainable living conditions.
- Funding from the Global Environment Facility (GEF) for countries with few financial resources.
- Inclusion of a for assessing ecosystem status.

Environmental Conventions are Regularly Criticized for Their:

- Rigidity and verticality: they are too descriptive, homogenous and top down, not reflecting the diversity and complexity of environmental issues. Signatory countries struggle to translate objectives into concrete form and incorporate them consistently.
- Duplicate structures and aid: the sector-specific format of the conventions produced duplicate structures and procedures. Inadequate cooperation between government ministries.
- Contradictions and incompatibility: e.g., "if reforestation projects to reduce CO_2 give preference to monocultures of exotic species, this can have a negative impact on biodiversity (whereas natural regeneration can strengthen both biodiversity and the conditions needed for life)."

Until now, the formulation of environmental policies at the international level has been divided by theme, sector or territory, resulting in treaties that overlap or clash. International attempts to coordinate environment institutions, include the Inter-Agency Coordination Committee and the Commission for Sustainable Development, but these institutions are not powerful enough to effectively incorporate the three aspects of sustainable development.

Multilateral Environmental Agreements (MEAs)

MEAs are agreements between several countries that apply internationally or regionally and concern a variety of environmental questions. As of 2013 over 500 Multilateral Environmental Agreements (MEAs), including 45 of global scope involve at least 72 signatory countries. Further agreements cover regional environmental problems, such as deforestation in Borneo or pollution in the Mediterranean. Each agreement has a specific mission and objectives ratified by multiple states.

Many Multilateral Environmental Agreements have been negotiated with the support from the United Nations Environmental Programme and work towards the achievement of the United Nations Millennium Development Goals as a means to instil sustainable practices for the environment and its people. Multilateral Environmental Agreements are considered to present enormous opportunities for greener societies and economies which can deliver numerous benefits in addressing food, energy and water security and in achieving sustainable development. These agreements can be implemented on a global or regional scale, for example the issues surrounding the disposal of hazardous waste can be implemented on a regional level as per the Bamako Convention on the Ban of the Import into Africa and the Control of Transboundary Movement and Management of Hazardous Waste within Africa which applies specifically to Africa, or the global approach to hazardous waste such as the Basel Convention on the Control of Transboundary Movements of Hazardous Wastes and their Disposal which is monitored throughout the world.

"The environmental governance structure defined by the Rio and Johannesburg Summits is sustained by UNEP, MEAs and developmental organizations and consists of assessment and policy development, as well as project implementation at the country level.

"The Governance Structure Consists of a Chain of Phases:

- Assessment of environment status;
- International policy development;
- Formulation of MEAs;
- Policy implementation;
- Policy assessment;
- Enforcement;
- Sustainable development.

"Traditionally, UNEP has focused on the normative role of engagement in the first three phases. Phases (d) to (f) are covered by MEAs and the sustainable development phase involves developmental organizations such as UNDP and the World Bank."

Lack of coordination affects the development of coherent governance. The report shows that donor states support development organizations, according to their individual interests. They do not follow a joint plan, resulting in overlaps and duplication. MEAs tend not to become a joint frame of reference and therefore receive little financial support. States and organizations emphasize existing regulations rather than improving and adapting them.

Background

The risks associated with nuclear fission raised global awareness of environmental threats. The 1963 Partial Nuclear Test Ban Treaty prohibiting atmospheric nuclear testing was the beginning of the globalization of environmental issues. Environmental law began to be modernized and coordinated with the Stockholm Conference (1972), backed up in 1980 by the Vienna Convention on the Law of Treaties. The Vienna Convention for the Protection of the Ozone Layer was signed and ratified in 1985. In 1987, 24 countries signed the Montreal Protocol which imposed the gradual withdrawal of CFCs.

The Brundtland Report, published in 1987 by the UN Commission on Environment and Development, stipulated the need for economic development that "meets the needs of the present without compromising the capacity of future generations to meet their needs.

Rio Conference (1992) and Reactions

The United Nations Conference on Environment and Development (UNCED), better known as the 1992 Earth Summit, was the first major international meeting since the end of the Cold War and was attended by delegations from 175 countries. Since then the biggest international conferences that take place every 10 years guided the global governance process with a series of MEAs. Environmental treaties are applied with the help of secretariats.

Governments created international treaties in the 1990s to check global threats to the environment. These treaties are far more restrictive than global protocols and set out to change non-sustainable production and consumption models.

Agenda 21

Agenda 21 is a detailed plan of actions to be implemented at the global, national and local levels by UN organizations, member states and key individual groups in all regions. Agenda 21 advocates making sustainable development a legal principle law. At the local level, local Agenda 21 advocates an inclusive, territory-based strategic plan, incorporating sustainable environmental and social policies.

The Agenda has been accused of using neoliberal principles, including free trade to achieve environmental goals. For example, chapter two, entitled "International Cooperation to Accelerate Sustainable Development in Developing Countries and Related Domestic Policies" states, "The international economy should provide a supportive international climate for achieving environment and development goals by: promoting sustainable development through trade liberalization."

Actors

International Institutions

United Nations Environment Program

The UNEP has had its biggest impact as a monitoring and advisory body, and in developing environmental agreements. It has also contributed to strengthening the institutional capacity of environment ministries.

In 2002 UNEP held a conference to focus on product lifecycle impacts, emphasizing the fashion, advertising, financial and retail industries, seen as key agents in promoting sustainable consumption.

According to Ivanova, UNEP adds value in environmental monitoring, scientific assessment and information sharing, but cannot lead all environmental management processes. She proposed the following tasks for UNEP:

- Initiate a strategic independent overhaul of its mission;
- Consolidate the financial information and transparency process;
- Restructure organizing governance by creating an operative executive council that balances the omnipresence of the overly imposing and fairly ineffectual Governing Council/Global Ministerial Environment Forum (GMEF).

Other proposals offer a new mandate to "produce greater unity amongst social and environmental agencies, so that the concept of 'environment for development' becomes a reality. It needs to act as a platform for establishing standards and for other types of interaction with national and international organizations and the United Nations. The principles of cooperation and common but differentiated responsibilities should be reflected in the application of this revised mandate."

Sherman proposed principles to strengthen UNEP:

- Obtain a social consensus on a long-term vision;
- Analyze the current situation and future scenarios;
- Produce a comprehensive plan covering all aspects of sustainable development;
- Build on existing strategies and processes;
- Multiply links between national and local strategies;
- Include all these points in the financial and budget plan;
- Adopt fast controls to improve process piloting and identification of progress made;
- Implement effective participation mechanisms.

Another group stated, "Consider the specific needs of developing countries and respect of the fundamental principle of 'common but differentiated responsibilities'. Developed countries should promote technology transfer, new and additional financial resources, and capacity building for meaningful participation of developing countries in international environmental governance. Strengthening of international environmental governance should occur in the context of sustainable development and should involve civil society as an important stakeholder and agent of transformation."

Global Environment Facility (GEF)

Created in 1991, the Global Environment Facility is an independent financial organization initiated by donor governments including Germany and France. It was the first financial organization dedicated to the environment at the global level. As of 2013 it had 179 members. Donations are used for projects covering biodiversity, climate change, international waters, destruction of the ozone layer, soil degradation and persistent organic pollutants.

GEF's institutional structure includes UNEP, UNDP and the World Bank. It is the funding mechanism for the four environmental conventions: climate change, biodiversity, persistent organic pollutants and desertification. GEF transfers resources from developed countries to developing countries to fund UNDP, UNEP and World Bank projects. The World Bank manages the annual budget of US$561.10 million.

The GEF has been criticized for its historic links with the World Bank, at least during its first phase during the 1990s, and for having favoured certain regions to the detriment of others. Another view sees it as contributing to the emergence of a global "green market". It represents "an adaptation (of the World Bank) to this emerging world order, as a response to the emergence of environmental movements that are becoming a geopolitical force." Developing countries demanded financial transfers to help them protect their environment.

GEF is subject to economic profitability criteria, as is the case for all the conventions. It received more funds in its first three years than the UNEP has since its creation in 1972. GEF funding represents less than 1% of development aid between 1992 and 2002.

United Nations Commission on Sustainable Development (CSD)

This intergovernmental institution meets twice a year to assess follow-up on Rio Summit goals. The CSD is made up of 53 member states, elected every three years and was reformed in 2004 to help improve implementation of Agenda 21. It meets twice a year, focusing on a specific theme during each two-year period: 2004-2005 was dedicated to water and 2006-2007 to climate change. The CSD has been criticized for its low impact, general lack of presence and the absence of Agenda 21 at the state level specifically, according to a report by the World Resources Institute. Its mission focuses on sequencing actions and establishing agreements puts it in conflict with institutions such as UNEP and OECD.

World Environment Organization (WEO)

A proposed World Environment Organization, analogous to the World Health Organization could be capable of adapting treaties and enforcing international standards.

The European Union, particularly France and Germany, and a number of NGOs favour creating a WEO. The United Kingdom, the US and most developing countries prefer to focus on voluntary initiatives. WEO partisans maintain that it could offer better political leadership, improved legitimacy and more efficient coordination. Its detractors argue that existing institutions and missions already provide appropriate environmental governance; however the lack of coherence and coordination between them and the absence of clear division of responsibilities prevents them from greater effectiveness.

World Bank

The World Bank influences environmental governance through other actors, particularly the GEF. The World Bank's mandate is not sufficiently defined in terms of environmental governance despite the fact that it is included in its mission. However, it allocates 5 to 10% of its annual funds to environmental projects. The institution's capitalist vocation means that its investment is concentrated solely in areas which are profitable in terms of cost benefits, such as climate change action and ozone layer protection, whilst neglecting other such as adapting to climate change and desertification. Its financial autonomy means that it can make its influence felt indirectly on the creation of standards, and on international and regional negotiations.

Following intense criticism in the 1980s for its support for destructive projects which, amongst other consequences, caused deforestation of tropical forests, the World Bank drew up its own environment-related standards in the 1990s so it could correct its actions. These standards differ from UNEP's standards, meant to be the benchmark, thus discrediting the institution and sowing disorder and conflict in the world of environmental governance. Other financial institutions, regional development banks and the private sector also drew up their own standards. Criticism is not directed at the World Bank's standards in themselves, which Najam considered as "robust", but at their legitimacy and efficacy.

GEF

The GEF's account of itself as of 2012 is as "the largest public funder of projects to improve the global environment", period, which "provides grants for projects related to biodiversity, climate change, international waters, land degradation, the ozone layer, and persistent organic pollutants." It claims to have provided "$10.5 billion in grants and leveraging $51 billion in co-financing for over 2,700 projects in over 165 countries [and] made more than 14,000 small grants directly to civil society and community-based organizations, totaling $634 million." It serves as mechanism for the:

- Convention on Biological Diversity (CBD).
- United Nations Framework Convention on Climate Change (UNFCCC).
- Stockholm Convention on Persistent Organic Pollutants (POPs).
- Convention to Combat Desertification (UNCCD).
- Implementation of Montreal Protocol on Substances That Deplete the Ozone Layer in some countries with "economies in transition".

This mandate reflects the restructured GEF as of October 2011.

World Trade Organization (WTO)

The WTO's mandate does not include a specific principle on the environment. All the problems linked to the environment are treated in such a way as to give priority to trade requirements and the principles of the WTO's own trade system. This produces conflictual situations. Even if the WTO recognizes the existence of MEAs, it denounces the fact that around 20 MEAs are in conflict with the WTO's trade regulations. Furthermore, certain MEAs can allow a country to ban or limit trade in certain products if they do not satisfy established environmental protection requirements. In these circumstances, if one country's ban relating to another country concerns two signatories of the same MEA, the principles of the treaty can be used to resolve the disagreement, whereas if the country affected by the trade ban with another country has not signed the agreement, the WTO demands that the dispute be resolved using the WTO's trade principles, in other words, without taking into account the environmental consequences.

Some criticisms of the WTO mechanisms may be too broad. In a recently dispute over labelling of dolphin safe labels for tuna between the US and Mexico, the ruling was relatively narrow and did not, as some critics claimed.

International Monetary Fund (IMF)

The IMF's mission is to encourage growth and development. The IMF advocates reduced public expenditure, increased exports and foreign investment. The environment is not a priority for the IMF, leading it to favor projects that may have negative environmental effects.

The IMF Green Fund proposal of Dominique Strauss-Kahn specifically to address "climate-related shocks in Africa", despite receiving serious attention was rejected. Strauss-Kahn's proposal, backed by France and Britain, was that "developed countries would make an initial capital injection into the fund using some of the $176 billion worth of SDR allocations from last year in exchange for a stake in the green fund." However, "most of the 24 directors ... told Strauss-Kahn that climate was not part of the IMF's mandate and that SDR allocations are a reserve asset never intended for development issues."

UN ICLEI

The UN's main body for coordinating municipal and urban decision-making is named the International Council for Local Environmental Initiatives. Its slogan is "Local Governments for Sustainability". This body sponsored the concept of full cost accounting that makes environmental governance the foundation of other governance.

ICLEIs projects and achievements include:

- Convincing thousands of municipal leaders to sign the World Mayors and Municipal Leaders Declaration on Climate Change (2005) which notably requests of other levels of government that.
- Global trade regimes, credits and banking reserve rules be reformed to advance debt relief and incentives to implement polices and practices that reduce and mitigate climate change.
- Starting national councils to implement this and other key agreements, e.g., ICLEI Local Governments for Sustainability USA.
- Spreading ecoBudget (2008) and Triple Bottom Line (2007) "tools for embedding sustainability into council operations", e.g. Guntur's Municipal Corporation, one of the first four to ipmlement the entire framework.
- Sustainability Planning Toolkit (launched 2009) integrating these and other tools.
- Cities Climate Registry (launched 2010) - part of UNEP Campaign on Cities and Climate Change.

ICLEI promotes best practice exchange among municipal governments globally, especially green infrastructure, sustainable procurement.

Other Secretariats

Other international institutions incorporate environmental governance in their action plans, including:

- United Nations Development Programme (UNDP), promoting development.
- World Meteorological Organization (WMO) which works on the climate and atmosphere.
- Food and Agriculture Organisation (FAO) working on the protection of agriculture, forests and fishing.
- International Atomic Energy Agency (IAEA) which focuses on nuclear security.

Over 30 UN agencies and programmes support environmental management, according to Najam. This produces a lack of coordination, insufficient exchange of information and dispersion of responsibilities. It also results in proliferation of initiatives and rivalry between them.

Criticism

According to Bauer, Busch and Siebenhüner, the different conventions and multilateral agreements of global environmental regulation is increasing their secretariats' influence. Influence varies according to bureaucratic and leadership efficiency, choice of technical or client-centered.

The United Nations is often the target of criticism, including from within over the multiplication of secretariats due to the chaos it produces. Using a separate secretariat for each MEA creates enormous overhead given the 45 international-scale and over 500 other agreements.

States

Environmental Governance at the State Level

Environmental protection has created opportunities for mutual and collective monitoring among neighbouring states. The European Union provides an example of the institutionalization of joint regional and state environmental governance. Key areas include information, led by the European Environment Agency (EEA), and the production and monitoring of norms by states or local institutions.

State Participation in Global Environmental Governance

US refusal to ratify major environment agreements produced tensions with ratifiers in Europe and Japan.

The World Bank, IMF and other institutions are dominated by the developed countries and do not always properly consider the requirements of developing countries.

Business

Environmental governance applies to business as well as government. Considerations are typical of those in other domains:

- Values (vision, mission, principles).
- Policy (strategy, objectives, targets).
- Oversight (responsibility, direction, training, communication).
- Process (management systems, initiatives, internal control, monitoring and review, stakeholder dialogue, transparency, environmental accounting, reporting and verification).
- Performance (performance indicators, benchmarking, eco-efficiency, reputation, compliance, liabilities, business development).

White and Klernan among others discuss the correlation between environmental governance and financial performance. This correlation is higher in sectors where environmental impacts are greater.

Business environmental issues include emissions, biodiversity, historical liabilities, product and material waste/recycling, energy use/supply and many others.

Environmental governance has become linked to traditional corporate governance as an increasing number of shareholders are corporate environmental impacts. Corporate governance is the set of processes, customs, policies, laws, and institutions affecting the way a corporation (or company) is managed. Corporate governance is affected by the relationships among stakeholders. These stakeholders research and quantify performance to compare and contrast the environmental performance of thousands of companies.

Large corporations with global supply chains evaluate the environmental performance of business partners and suppliers for marketing and ethical reasons. Some consumers seek environmentally friendly and sustainable products and companies.

Non-Governmental Organizations

According to Bäckstrand and Saward, "broader participation by non-state actors in multilateral environmental decisions (in varied roles such as agenda setting, campaigning, lobbying, consultation, monitoring, and implementation) enhances the democratic legitimacy of environmental governance."

Local activism is capable of gaining the support of the people and authorities to combat environmental degradatation. In Cotacachi, Ecuador, a social movement used a combination of education, direct action, the influence of local public authorities and denunciation of the mining company's plans in its own country, Canada, and the support of international environmental groups to influence mining activity.

Fisher cites cases in which multiple strategies were used to effect change. She describes civil society groups that pressure international institutions and also organize local events. Local groups can take responsibility for environmental governance in place of governments.

According to Bengoa, "social movements have contributed decisively to the creation of an institutional platform wherein the fight against poverty and exclusion has become an inescapable benchmark." But despite successes in this area, "these institutional changes have not produced the processes for transformation that could have made substantial changes to the opportunities available to rural inhabitants, particularly the poorest and those excluded from society." He cites several reasons:

- Conflict between in-group cohesion and openness to outside influence.
- Limited trust between individuals.
- Contradiction between social participation and innovation.
- Criticisms without credible alternatives to environmentally damaging activities.

A successful initiative in Ecuador involved the establishment of stakeholder federations and management committees (NGOs, communities, municipalities and the ministry) for the management of a protected forest.

Proposals

The International Institute for Sustainable Development proposed an agenda for global governance. These objectives are:

- Expert leadership.
- Positioning science as the authoritative basis of sound environmental policy.
- Coherence and reasonable coordination.
- Well-managed institutions.
- Incorporate environmental concerns and actions within other areas of international policy and action.

Coherence and Coordination

Despite the increase in efforts, actors, agreements and treaties, the global environment continue to degrade at a rapid rate. From the big hole in Earth's ozone layer to over-fishing to the uncertainties of climate change, the world is confronted by several intrinsically global challenges. However, as the environmental agenda becomes more complicated and extensive, the current system has proven ineffective in addressing and tackling problems related to trans-boundary externalities and the environment is still experiencing degradation at unprecedented levels.

Inforesources identifies four major obstacles to global environmental governance, and describes measures in response. The four obstacles are:

- Parallel structures and competition, without a coherent strategy.
- Contradictions and incompatibilities, without appropriate compromise.
- Competition between multiple agreements with incompatible objectives, regulations and processes.
- Integrating policy from macro- to micro- scales.

Recommended Measures

- MDGs (Milenium Development Goals) and conventions, combining sustainability and reduction of poverty and equity.
- Country-level approach linking global and local scales.
- Coordination and division of tasks in a multilateral approach that supports developing countries and improves coordination between donor countries and institutions.
- Use of Poverty Reduction Strategy Papers (PRSPs) in development planning.
- Transform conflicts into tradeoffs, synergies and win-win options.

Contemporary debates surrounding global environmental governance have converged on the idea of developing a stronger and more effective institutional framework. The views on how to achieve this, however, still hotly debated. Currently, rather than teaming up with the United Nations Environment Programme (UNEP), international environmental responsibilities have been spread across many different agencies including: a) specialised agencies within the UN system such as the World Meteorological Organisation, the International Maritime Organisation and others; b) the programs in the UN system such as the UN Development Program; c) the UN regional economic and social commission; d) the Bretton Woods institutions; e) the World Trade Organisation and; f) the environmentally focused mechanisms such as the Global Environment Facility and close to 500 international environmental agreements.

Some analysts also argue that multiple institutions and some degree of overlap and duplication in policies is necessary to ensure maximum output from the system. Others, however, claim that institutions have become too dispersed and lacking in coordination which can be damaging to their effectiveness in global environmental governance. Whilst there are various arguments for and against a WEO, the key challenge, however, remains the same: how to develop a rational and effective framework that will protect the global environment efficiently.

Democratization

Starting in 2002, Saward and others began to view the Earth Summit process as capable opening up the possibility of stakeholder democracy. The summits were deliberative rather than simply participative, with NGOs, women, men, indigenous peoples and businesses joining the decision-making process alongside states and international organizations, characterized by:

- The importance given to scientific and technical considerations.
- The official and unofficial participation of many actors with heterogeneous activity scopes
- Growing uncertainty.
- A new interpretation of international law and social organization models.

As of 2013, the absence of joint rules for composing such fora leads to the development of non-transparent relations that favour the more powerful stakeholders. Criticisms assert that they act more as a lobbying platform, wherein specific interest groups attempt to influence governments.

Institutional Reform

Actors inside and outside the United Nations are discussing possibilities for global environmental governance that provides a solution to current problems of fragility, coordination and coherence. Deliberation is focusing on the goal of making UNEP more efficient. A 2005 resolution recognizes "the need for more efficient environmental activities in the United Nations system, with enhanced coordination, improved policy advice and guidance, strengthened scientific knowledge, assessment and cooperation, better treaty compliance, while respecting the legal autonomy of the treaties, and better integration of environmental activities in the broader sustainable development framework."

Proposals Include

- Greater and better coordination between agencies.
- Strengthen and acknowledge UNEP's scientific role.
- Identify MEA areas to strengthen coordination, cooperation and teamwork between different agreements.
- Increase regional presence.
- Implement the Bali Strategic Plan on improving technology training and support for the application of environmental measures in poor countries.
- Demand that UNEP and MEAs participate formally in all relevant WTO committees as observers.
- Strengthen its financial situation.
- Improve secretariats' efficiency and effectiveness.

One of the Main Studies Addressing this Issue Proposes

- Clearly divide tasks between development organizations, UNEP and the MEAs.
- Adopt a political direction for environmental protection and sustainable development.
- Authorize the UNEP Governing Council/Global Ministerial Environment Forum to adopt the UNEP medium-term strategy.
- Allow Member States to formulate and administer MEAs an independent secretariat for each convention.
- Support UNEP in periodically assessing MEAs and ensure coordination and coherence.
- Establish directives for setting up national/regional platforms capable of incorporating MEAs in the Common Country Assessment (CCA) process and United Nations Development Assistance Framework (UNDAF).
- Establish a global joint planning framework.
- Study the aptitude and efficiency of environmental activities' funding, focusing on differential costs.
- Examine and redefine the concept of funding differential costs as applicable to existing financial mechanisms.
- Reconsider remits, division of tasks and responsibilities between entities that provide services to the multipartite conferences. Clearly define the services that UN offices provide to MEA secretariats.
- propose measures aiming to improve personnel provision and geographic distribution for

MEA secretariats.

- Improve transparency resource use for supporting programmes and in providing services to MEAs. Draw up a joint budget for services supplied to MEAs.

Education

A 2001 Alliance 21 report proposes six fields of action:

- Strengthen citizens' critical faculties to ensure greater democratic control of political orientations.
- Develop a global and critical approach.
- Develop civic education training for teachers.
- Develop training for certain socio-professional groups.
- Develop environmental education for the entire population.
- Assess the resulting experiences of civil society.

Transform Daily Life

Individuals can modify consumption, based on voluntary simplicity: changes in purchasing habits, simplified lifestyles (less work, less consumption, more socialization and constructive leisure time). But individual actions must not replace vigilance and pressure on policies. Notions of responsible consumption developed over decades, revealing the political nature of individual purchases, according to the principle that consumption should satisfy the population's basic needs. These needs comprise the physical wellbeing of individuals and society, a healthy diet, access to drinking water and plumbing, education, healthcare and physical safety. The general attitude centres on the need to reduce consumption and reuse and recycle materials. In the case of food consumption, local, organic and fair trade products which avoid ill treatment of animals has become a major trend.

Alternatives to the personal automobile are increasing, including public transport, car sharing and bicycles and alternative propulsion systems.

Alternative energy sources are becoming less costly.

Ecological industrial processes turn the waste from one industry into raw materials for another.

Governments can reduce subsidies/increase taxes/tighten regulation on unsustainable activities.

The Community Environmental governance Global Alliance encourages holistic approaches to environmental and economic challenges, incorporating indigenous knowledge. Okotoks, Alberta capped population growth based on the carrying capacity of the Sheep River. The Fraser Basin Council Watershed Governance in British Columbia, Canada, manages issues that span municipal jurisdictions. Smart Growth is an international movement that employs key tenets of Environmental governance in urban planning.

Policies and Regulations

Establish policies and regulations that promote "infrastructures for well being" whilst addressing the political, physical and cultural levels.

Eliminate subsidies that have a negative environmental impact and tax pollution.

Promoting workers' personal and family development.

Coordination

A programme of national workshops on synergies between the three Rio Conventions launched in late 2000, in collaboration with the relevant secretariats. The goal was to strengthen coordination at the local level by:

- Sharing information.
- promoting political dialogue to obtain financial support and implement programmes.
- Enabling the secretariats to update their joint work programmes.

According to Campbell, "In the context of globalization, the question of linking up environmental themes with other subjects, such as trade, investment and conflict resolution mechanisms, as well as the economic incentives to participate in and apply agreements would seem to provide an important lesson for the effective development of environmental governance structures." Environmental concerns would become part of the global economic system. "These problems also contain the seeds of a new generation of international conflicts that could affect both the stability of international relations and collective security. Which is why the concept of 'collective security' has arisen."

Moving local decisions to the global level is as important as the way in which local initiatives and best practices are part of a global system. Kanie points out that NGOs, scientists, international institutions and stakeholder partnerships can reduce the distance that separates the local and international levels.

References

- *Rushefsky, Mark E. (2002). Public Policy in the United States at the Dawn of the Twenty-first Century (3rd ed.). New York: M.E. Sharpe, Inc. pp. 253–254. ISBN 978-0-7656-1663-0*
- Hardman Reis, T., *Compensation for Environmental Damages Under International Law*, Kluwer Law International, The Hague, 2011, ISBN 978-90-411-3437-0
- *Chapman, Roger (2010). Culture wars: an encyclopedia of issues, viewpoints, and voices. M.E. Sharpe, Inc. p. 162. ISBN 0-7656-1761-7*
- *Greg Barton (2002). Empire Forestry and the Origins of Environmentalism. Cambridge University Press. p. 48. ISBN 9781139434607*
- *Steve Yearly (15 April 2008). "Globalization and the Environment". In George Ritzer. The Blackwell Companion to Globalization. John Wiley & Sons. p. 240. ISBN 978-0-470-76642-2*
- *Miller, Jr., G. Tyler (2003). Environmental Science: Working With the Earth (9th ed.). Pacific Grove, California: Brooks/Cole. p. G5. ISBN 0-534-42039-7*
- *Wu and Wen (2015). Nongovernmental organizations and environmental protests: Impacts in East Asia*

(chapter 7 of Routledge Handbook of Environment and Society in Asia). London: Routledge. pp. 105–119. ISBN 978-0-415-65985-7

- *Betty Dobratz; Lisa K Waldner; Timothy Buzzell (14 October 2015). Power, Politics, and Society: An Introduction to Political Sociology. Routledge. p. 346. ISBN 978-1-317-34529-9*
- *John Benyon; David Dunkerley (1 May 2014). Globalization: The Reader. Routledge. p. 54. ISBN 978-1-136-78240-4*
- *Grainger, Alan (2012-01-01). Environmental Globalization. John Wiley & Sons, Ltd. doi:10.1002/9780470670590.wbeog170/full. ISBN 9780470670590*
- *Karl S. Zimmerer (2006). Globalization & New Geographies of Conservation. University of Chicago Press. p. 1. ISBN 978-0-226-98344-8*
- *Alan Grainger (31 October 2013). "Environmental Globalization and Tropical Forests". In Jan Oosthoek; Barry K. Gills. The Globalization of Environmental Crisis. Routledge. p. 63. ISBN 978-1-317-96896-2*
- *Eccleston, Charles; Doub, J. Peyton (2012). Preparing NEPA Environmental Assessments: A User's Guide to Best Professional Practices. CRC Press. ISBN 9781439808825*

5

Environmental Management Systems

The database and system which incorporates processes and procedures to train personnel, monitor, summarize and report specialized environmental performance information to the external as well as the internal stakeholders of an organization is known as an environmental management system. This chapter discusses the different aspects of environmental management systems in detail.

Environmental management system (EMS) refers to the management of an organization's environmental programs in a comprehensive, systematic, planned and documented manner. It includes the organizational structure, planning and resources for developing, implementing and maintaining policy for environmental protection.

More formally, EMS is "a system and database which integrates procedures and processes for training of personnel, monitoring, summarizing, and reporting of specialized environmental performance information to internal and external stakeholders of a firm."

The most widely used standard on which an EMS is based is International Organization for Standardization (ISO) 14001. Alternatives include the EMAS.

An environmental management information system (EMIS) is an information technology solution for tracking environmental data for a company as part of their overall environmental management system.

Goals

The goals of EMS are to increase compliance and reduce waste:

- Compliance is the act of reaching and maintaining minimal legal standards. By not being compliant, companies may face fines, government intervention or may not be able to operate.
- Waste reduction goes beyond compliance to reduce environmental impact. The EMS helps to develop, implement, manage, coordinate and monitor environmental policies. Waste reduction begins at the design phase through pollution prevention and waste minimization. At the end of the life cycle, waste is reduced by recycling.

To meet these goals, the selection of environmental management systems is typically subject to a certain set of criteria: a proven capability to handle high frequency data, high performance indicators, transparent handling and processing of data, powerful calculation engine, customised factor handling, multiple integration capabilities, automation of workflows and QA processes and in-depth, flexible reporting.

Features

An environmental management system (EMS):

- Serves as a tool, or process, to improve environmental performance and information mainly "design, pollution control and waste minimization, training, reporting to top management, and the setting of goals".
- Provides a systematic way of managing an organization's environmental affairs.
- Is the aspect of the organization's overall management structure that addresses immediate and long-term impacts of its products, services and processes on the environment. EMS assists with planning, controlling and monitoring policies in an organization.
- Gives order and consistency for organizations to address environmental concerns through the allocation of resources, assignment of responsibility and ongoing evaluation of practices, procedures and processes.
- Creates environmental buy-in from management and employees and assigns accountability and responsibility.
- Sets framework for training to achieve objectives and desired performance.
- Helps understand legislative requirements to better determine a product or service's impact, significance, priorities and objectives.
- Focuses on continual improvement of the system and a way to implement policies and objectives to meet a desired result. This also helps with reviewing and auditing the EMS to find future opportunities.
- Encourages contractors and suppliers to establish their own EMS.

EMS Model

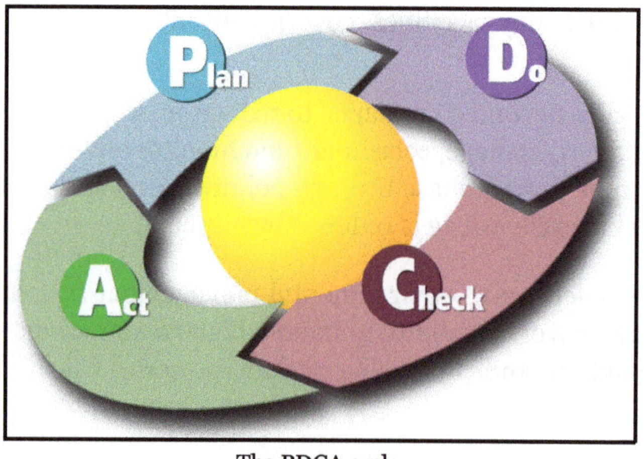

The PDCA cycle.

An EMS follows a Plan-Do-Check-Act, or PDCA, Cycle. The diagram shows the process of first developing an environmental policy, planning the EMS, and then implementing it. The process also includes checking the system and acting on it. The model is continuous because an EMS is a process of continual improvement in which an organization is constantly reviewing and revising the system.

This is a model that can be used by a wide range of organizations — from manufacturing facilities to service industries to government agencies.

Other Meanings

An EMS can also be classified as:

- A system which monitors, tracks and reports emissions information, particularly with respect to the oil and gas industry. EMSs are becoming web-based in response to the EPA's mandated greenhouse gas (GHG) reporting rule, which allows for reporting GHG emissions information via the internet.
- A centrally controlled and often automated network of devices (now frequently wireless using z-wave and zigbee technologies) used to control the internal environment of a building. Such a system namely acts as an interface between end user and energy (gas/electricity) consumption.

Examples of Environmental Management Systems

- Emisoft's environmental management, reporting and compliance platform.
- Medgate environmental management software.
- EsDat environmental data management system.
- Enviance regulatory compliance system.

The growing public interest in environmental issues and concern for environmental quality has led to the emergence of strict pollution control regimes. This has brought about the development and implementation of various voluntary schemes and standards for environmental management and pollution control. Notable examples are BS 7750, the European Union's Eco-Management and Audit Scheme (EMAS) and the ISO 14000 series. In this context, putting in place systems to meet with these standards assume significance. An environmental management system (EMS) of an organisation is a system designed to:

- Meet the regulatory and legislative system requirements.
- Improve the control of the environmental impact.
- Provide confidence to the customers that the products and services are manufactured with the aim of reducing the negative impact on the environment.
- Suitably accommodate changing market trends and gain competitive edge.
- Reduce the costs associated with environmental liabilities and insurance.
- Gain public and media support.

EMS involves preparing a list of all environmental legislation, making a register of all environmental effects associated with the company's activities, setting targets (such as reducing waste production by a quantifiable amount), keeping appropriate records and undertaking regular reviews of the system. The whole system, in essence, forms the basis for sound environmental performance.

Most EMSs are built on the notational model. This 'plan-do-check-act' model leads to continuous improvement based upon:

- Planning, including identifying environmental aspects and establishing goals (plan);
- Implementing, including training and operational controls (do);
- Checking, including monitoring and corrective action (check);
- Reviewing, including progress reviews and acting to incorporate required changes in the EMS (act).

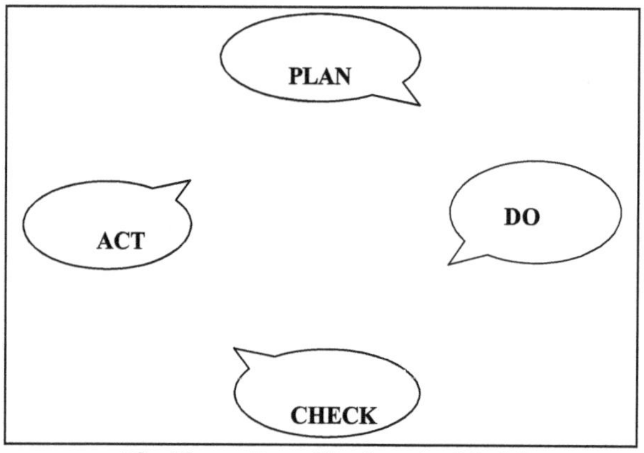

The Plan – Do – Check – Act Model.

To reiterate, this model ensures that environmental matters are systematically identified, controlled and monitored. It also helps ensure that performance of the EMS is improved over time.

Core Elements of EMS

There is no single best approach to the development and implementation of EMS, since it depends on the nature, size and complexity of the activities, products and services within the organisation. All environmental management systems, however, have a number of core elements in common and these include the following:

- Environmental policy: An environmental policy is usually published as a written statement, expressing the commitment of the senior management to improving appropriate environmental performance. It is most often understood as a public statement of the intentions and principles of action for the organisation regarding the environment. The policy statement should define the broad environmental goals the organisation has decided to achieve. These are most clear if they are quantified, e.g., to reduce emissions of pollutants by 95% within 5 years, to provide sewerage and biological treatment of sludge for 60% of the population within 3 years, etc.

- Environment action plan: An environment programme or action plan describes the measures the organisation will take over the coming year(s). The action plan essentially translates the environmental policies of the organisation into objectives and targets and identifies the activities to achieve them, defines responsibilities and commits the necessary human and financial resources for implementation. This includes committing the necessary funds and staff to meet each goal and providing for monitoring and co-ordination of the progress made towards achieving these separate goals and the overall policy goals that have to be fulfilled.
- Environment related organisational structures: The structures establish assignments, delegate authority and assign responsibility for actions. In the case of organisations with multiple sites or different activities, structures for the organisation as a whole as well as for the separate operating units are necessary. The senior staff member responsible for the environment, typically, has a direct reporting relationship to the head of the organisation. Individuals holding strategic or main environmental responsibilities should be adequately supported with human and financial resources.
- Integration of environmental concerns: The integration of environmental management into regular operation means the establishment of procedures for incorporating environmental measures into other operational aspects such as the protection of workers, purchasing, R&D, product development and acquisitions, marketing, finance, etc., in the case of companies and the safety, health and welfare of the community in the case of a local government. This encompasses the development of specific environmental procedures, usually detailed in operating manuals and other operating instructions describing measures and actions to take in the implementation of the environmental programme or action plan.

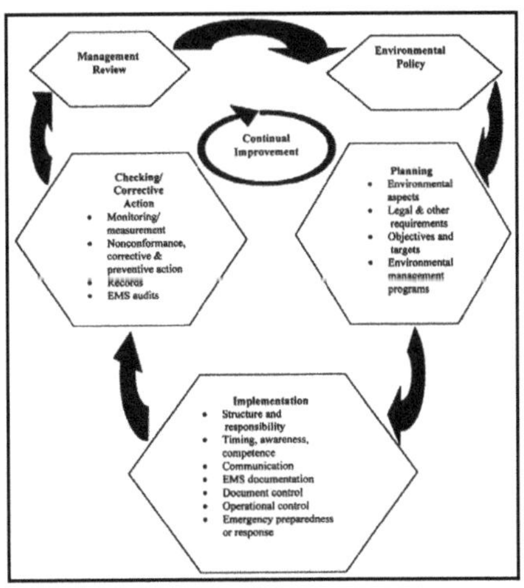

Because of the cyclic nature of the model, EMS is very dynamic in that a change or revision in any of the core elements of the system will have a chain/cascading effect on other elements.

Note that while there are several EMS models available, the model presented above uses the standards set by ISO 14001 as a starting point for describing EMS elements. This is so, mainly, due to the fact that.

- ISO 14001 is a widely accepted international standard for EMS that focuses on continual improvement.
- Companies may be asked to demonstrate conformance with ISO 14001 as a condition of doing business in some markets.

Benefits of EMS

A number of benefits are associated with adoption of EMS. For example, insurance companies are beginning to exclude pollution incidents from their insurance cover, and when they are included, the insurance companies place a surcharge on the policies. In order to improve or reduce any increases in the insurance premiums, companies can carry out an environmental audit and implement a EMS.

From a managerial point of view, a EMS enables a company to develop and maintain a well-organised management structure that ensures compliance with environmental legislation. By keeping the working place cleaner and safer, and the production more efficient, staff interest can be aroused and morale improved. Another incentive is the potential financial gains, which can be achieved. Reducing wastes and emissions and improving their treatment helps many companies to save a considerable sum of money in the long term. Adopting a EMS will also facilitate trade, as major trading blocs are more likely to accept members with a certain level of environmental performance. In other words, the benefits of a EMS include:

- Improved environmental performance.
- Enhanced compliance.
- Prevention of pollution/resource conservation.
- New customers/markets.
- Increased efficiency/reduced costs.
- Enhanced employee morale.
- Enhanced image with public, regulators, lenders, investors.
- Employee awareness of environmental issues and responsibilities.

Certification Body Assessments of EMS

To be certified, an organisation has to demonstrate that the EMS functions and the various control mechanisms are properly operational. In practice, this means that the EMS has been operational for a minimum of 3 months, the internal audit system is fully operational and that one management review has been conducted (Tech Monitor, Vol. 12. No. 5 Sept-Oct 1995).

Based on the following conditions, certificates are issued.

Non-conformities identified have been addressed and eliminated.

The certification body has justified confidence that all provisions in the EMS standard have been met, and, in particular, that provision for compliance with the organisation's policy and objectives is effective.

The principle of continuous improvement of environmental performance has been stated within an environmental programme and is being adhered to play in the EMS.

All key staff (i.e., those involved in managing significant effects) have undergone a training need analysis and have received training.

Regulatory Compliance

An organisation with a certified EMS has a system that should achieve continuing compliance with regulatory requirements applicable to the environmental aspects of its activities, products and services. The certification body confirms that a system designed to achieve the necessary compliance is operating effectively. In situations where authorities issue licenses or permits, it will often be a requirement that the organisation notifies the authority of any breaches. The certification body should ensure that the organisation's EMS records any infringement of regulations and that the appropriate corrective action has been taken.

An organisation with a certified EMS is responsible for continuing compliance with all the regulatory requirements and, therefore, must maintain a record of incidents of non-compliance and the remedial action.

Surveillance Audits

At each surveillance, the certification body should pay attention to the following:

- The effectiveness of the EMS with regard to achieving the objectives of the organisation's environmental policy;
- An interview with management responsible for the EMS;
- The functioning of procedures for notifying the authorities of any breaches;
- progress of planned activities aimed at continuous improvement of environmental performance, where applicable;
- Follow up conclusions resulting from internal audits;
- Action taken on non-conformities identified during the previous audit.

The certification body should be able to adapt its surveillance programme to the environmental issues related to the activities of the organisation and justify this arrangement. The surveillance audit of the certification body should be agreed with the organisation, taking into account the internal audit programme and the reliability that can be attributed to this work.

Documentation for EMS

While data relates to fact, information refers to a series of data analysed to provide a decision. Information is an answer to a question. All information comes from data, but not all data come from information (Tech Monitor, Vol. 12. No. 5 Sept-Oct 1995). In other words, executives making decisions often receive excessive amounts of data, but they do not always get the information they need, presented clearly and adequately for use in decisionmaking.

Decisions are usually related to solving problems. In order to solve a problem, one needs information to decide. A EMS, therefore, requires a proper documentation system in order to collect, analyse, register and retrieve information.

The documentation should describe the EMS and make clear the relationship to any other management system in operation within the organisation, or as having an influence on the operations of the entity, subject to certification. It is acceptable to combine the documentation for environmental and other management systems (such as for quality of health and safety), as long as the EMS can be clearly identified, together with the appropriate interfaces between different EMS documentations and the order of precedence.

Environmental effects may have been considered within safety reviews, hazard and operability studies and risk/consequence analysis, etc. In such cases, the EMS documentation should refer to the critical areas where significant effects are covered by other management systems. The EMS documentation should be available to all appropriate staff and should be kept up-to-date.

Environmental Management Manual

A key EMS document is a company's environment manual. This is a document that establishes the general policy of a company on this issue. It usually contains an environmental policy and a clear statement on the person who is in charge of verification of activities in the organisational structure pertaining to the maintenance of the environmental system. It lays down how each requirement of the policy will be handled in the organisation and includes a list of all procedures.

The manual also contains a description of the system, which was or is being implemented. There are considerable benefits in preparing a manual:

- Clarifies responsibilities;
- Is useful for audit purposes;
- Can be highly effective for training of new employees;
- Makes easier the process of obtaining approval of licenses from protection agencies;
- Is useful for marketing purposes.

The highest executive in the company must sign the document. It should contain the vision of the organisation about environment and its commitment to continuous improvement. The document should detail targets to be reached and explain how they will be reached, with what resources and under whose responsibility. The company must also realise that it is necessary to measure results in order to evaluate how correctly the policy is being applied. The company should, therefore, have a programme for achieving its objectives and targets.

To prepare the manual, it is necessary first to identify all mandatory regulations, standards and codes to be complied with. In addition, the company needs to identify other requirements, such as headquarters' policies, top management's strategic vision, market demands, etc.

The discussions that we have had so far provide you with a basic knowledge of EMS. Now, we will next discuss the importance of ISO 14000 and its supporting systems.

ISO 14000

ISO 14000 is a family of standards related to environmental management that exists to help organizations (a) minimize how their operations (processes, etc.) negatively affect the environment (i.e., cause adverse changes to air, water, or land); (b) comply with applicable laws, regulations, and other environmentally oriented requirements.

The current version of ISO 14001 is ISO 14001:2015 which was published in September 2015.

ISO 14000 is similar to ISO 9000 quality management in that both pertain to the process of how a product is produced, rather than to the product itself. As with ISO 9001, certification is performed by third-party organizations rather than being awarded by ISO directly. The ISO 19011 ISO 17021 audit standard apply when auditins are being performed.

The requirements of ISO 14001 are an integral part of the European Union's Eco-Management and Audit Scheme (EMAS). EMAS's structure and material requirements are more demanding, mainly concerning performance improvement, legal compliance, and reporting duties.

Brief History of Environmental Management Systems

In 1992, BSI Group published the world's first environmental management systems standard, BS 7750. Prior to this, environmental management had been part of larger systems such as Responsible Care. BS 7750 supplied the template for the development of the ISO 14000 series in 1996, by the International Organization for Standardization, which has representation from committees all over the world (ISO) (Clements 1996, Brorson & Larsson, 1999). As of 2010, ISO 14001 is now used by at least 223 149 organizations in 159 countries and economies.

Development of the ISO 14000 Series

The ISO 14000 family includes most notably the ISO 14001 standard, which represents the core set of standards used by organizations for designing and implementing an effective Environmental Management System (EMS). Other standards included in this series are ISO 14004, which gives additional guidelines for a good EMS, and more specialized standards dealing with specific aspects of environmental management. The major objective of the ISO 14000 series of norms is "to promote more effective and efficient environmental management in organizations and to provide useful and usable tools--ones that are cost-effective, system-based, [and] flexible, and reflect the best organizations and the best organizational practices available for gathering, interpreting, and communicating environmentally relevant information".

ISO 14000 series is based on a voluntary approach to environmental regulation (Szymanski &

Tiwari 2004). The series includes the ISO 14001 standard, which provides guidelines for the establishment or improvement of an EMS. The standard shares many common traits with its predecessor, ISO 9000, the international standard of quality management (Jackson 1997), which served as a model for its internal structure (National Academy Press 1999), and both can be implemented side by side. As with ISO 9000, ISO 14000 acts both as an internal management tool and as a way of demonstrating a company's environmental commitment to its customers and clients (Boiral 2007).

Prior to the development of the ISO 14000 series, organizations voluntarily constructed their own EMSs, but this made comparisons of environmental effects between companies difficult; therefore, the universal ISO 14000 series was developed. An EMS is defined by ISO as: "part of the overall management system, that includes organizational structure, planning activities, responsibilities, practices, procedures, processes, and resources for developing, implementing, achieving, and maintaining the environmental policy" (ISO 1996 cited in Federal Facilities Council Report 1999).

ISO 14001 Standard

ISO 14001 sets out the criteria for an Environmental Management System (EMS). It does not state requirements for environmental performance, but maps out a framework that a company or organization can follow to set up an effective EMS. It can be used by any organization that wants to improve resource efficiency, reduce waste, and drive down costs. Using ISO 14001 can provide assurance to company management and employees as well as external stakeholders that environmental impact is being measured and improved. ISO 14001 can also be integrated with other management functions and assists companies in meeting their environmental and economic goals.

ISO 14001, as with other ISO 14000 standards, is voluntary (IISD 2010), with its main aim to assist companies in continually improving their environmental performance, while complying with any applicable legislation. Organizations are responsible for setting their own targets and performance measures, with the standard serving to assist them in meeting objectives and goals and in the subsequent monitoring and measurement of these (IISD 2010).

The standard can be applied to a variety of levels in the business, from organizational level, right down to the product and service level (RMIT university). Rather than focusing on exact measures and goals of environmental performance, the standard highlights what an organization needs to do to meet these goals (IISD 2010).

ISO 14001 is known as a generic management system standard, meaning that it is relevant to any organization seeking to improve and manage resources more effectively. This includes:

- Single-site to large multi-national companies.
- High-risk companies to low-risk service organizations.
- Manufacturing, process, and the service industries, including local governments.
- All industry sectors including public and private sectors.
- Original equipment manufacturers and their suppliers.

All standards are periodically reviewed by ISO to ensure they still meet market requirements. The current version ISO 14001:2004 was last reviewed in 2012. The ISO committee decided a revision was necessary. The new version is expected by the end of 2015. After the revision has been published, certified organizations get a three-year transition period to adapt their environmental management system to the new edition of the standard. The new version of ISO 14001 is going to focus on the improvement of environmental performance rather than to improve the management system itself.

The new ISO 14001:2015 standard has been published and includes several new updates all aimed at making environmental management more comprehensive and relevant to the supply chain. One of the main updates is to consider environmental impacts during the life cycle – although there is no requirement to actually complete a life cycle analysis. In addition the commitments of top management and the methods of evaluating compliance have also been strengthened.

Another significant change has been the linking of ISO 14001 to the general management system structure, introduced in 2015, called the High Level Structure. Both ISO 9001 and 14001 use this same structure making implementation and auditing more uniform. The new standard also requires the holder of the certificate to specify risks and opportunities and how to address them.

Plan: Establish Objectives and Processes Required

Prior to implementing ISO 14001, an initial review or gap analysis of the organization's processes and products is recommended, to assist in identifying all elements of the current operation and, if possible, future operations, that may interact with the environment, termed "environmental aspects" (Martin 1998). Environmental aspects can include both direct, such as those used during manufacturing, and indirect, such as raw materials (Martin 1998). This review assists the organization in establishing their environmental objectives, goals, and targets, which should ideally be measurable; helps with the development of control and management procedures and processes; and serves to highlight any relevant legal requirement, which can then be built into the policy (Standards Australia/Standards New Zealand 2004).

Do: Implement the Processes

During this stage, the organization identifies the resources required and works out those members of the organization responsible for the EMS' implementation and control (Martin 1998). This includes establishing procedures and processes, although only one documented procedure is specified related to operational control. Other procedures are required to foster better management control over elements such as documentation control, emergency preparedness and response, and the education of employees, to ensure that they can competently implement the necessary processes and record results (Standards Australia/Standards New Zealand 2004). Communication and participation across all levels of the organization, especially top management, is a vital part of the implementation phase, with the effectiveness of the EMS being dependent on active involvement from all employees.

Check: Measure and Monitor the Processes and Report Results

During the "check" stage, performance is monitored and periodically measured to ensure that the organization's environmental targets and objectives are being met (Martin 1998). In addition, internal audits are conducted at planned intervals to ascertain whether the EMS meets the user's expectations and whether the processes and procedures are being adequately maintained and monitored (Standards Australia/Standards New Zealand 2004).

Act: take Action to Improve Performance of EMS Based on Results

After the checking stage, a management review is conducted to ensure that the objectives of the EMS are being met, the extent to which they are being met, and that communications are being appropriately managed; and to evaluate changing circumstances, such as legal requirements, in order to make recommendations for further improvement of the system (Standards Australia/ Standards New Zealand 2004). These recommendations are incorporated through continual improvement: plans are renewed or new plans are made, and the EMS moves forward.

Continual Improvement Process (CI)

ISO 14001 encourages a company to continually improve its environmental performance. Apart from the obvious – the reduction in actual and possible negative environmental impacts – this is achieved in three ways:

- Expansion: More and more business areas get covered by the implemented EMS.
- Enrichment: More and more activities, products, processes, emissions, resources, etc. get managed by the implemented EMS.
- Upgrading: An improvement of the structural and organizational framework of the EMS, as well as an accumulation of know-how in dealing with business-environmental issues.

Overall, the CI concept expects the organization to gradually move away from merely operational environmental measures towards a strategic approach on how to deal with environmental challenges.

Benefits

ISO 14001 was developed primarily to assist companies with a framework for better management control that can result in reducing their environmental impacts. In addition to improvements in performance, organizations can reap a number of economic benefits including higher conformance with legislative and regulatory requirements (Sheldon 1997) by adopting the ISO standard. By minimizing the risk of regulatory and environmental liability fines and improving an organization's efficiency (Delmas 2009), benefits can include a reduction in waste, consumption of resources, and operating costs. Secondly, as an internationally recognized standard, businesses operating in multiple locations across the globe can leverage their conformance to ISO 14001, eliminating the need for multiple registrations or certifications (Hutchens 2010). Thirdly, there has been a push in the last decade by consumers for companies to adopt better internal controls, making the incorporation of ISO 14001 a smart approach for the long-term viability of businesses. This can provide them with a competitive advantage against companies that do not adopt the standard (Potoki & Prakash, 2005). This in turn can have a positive impact on a company's asset value (Van der Deldt, 1997). It can lead to improved public perceptions of the business, placing them in a better position to operate in the international marketplace (Potoki & Prakash 1997; Sheldon 1997). The use of ISO 14001 can demonstrate an innovative and forward-thinking approach to customers and prospective employees. It can increase a business's access to new customers and business partners. In some markets it can potentially reduce public liability insurance costs. It can serve to reduce trade barriers between registered businesses (Van der Deldt, 1997). There is growing interest in including certification to ISO 14001 in tenders for public-private partnerships for infrastructure renewal. Evidence of value in terms of environmental quality and benefit to the taxpayer has been shown in highway projects in Canada.

Conformity Assessment

ISO 14001 can be used in whole or in part to help an organization (for-profit or not-for-profit) better manage its relationship with the environment. If all the elements of ISO 14001 are incorporated into the management process, the organization may opt to prove that it has achieved full alignment or conformity with the international standard, ISO 14001, by using one of four recognized options. These are:

1. make a self-determination and self-declaration, or
2. seek confirmation of its conformance by parties having an interest in the organization, such as customers, or
3. seek confirmation of its self-declaration by a party external to the organization, or
4. seek certification/registration of its EMS by an external organization.

ISO does not control conformity assessment; its mandate is to develop and maintain standards. ISO has a neutral policy on conformity assessment. One option is not better than the next. Each option serves different market needs. The adopting organization decides which option is best for them, in conjunction with their market needs.

Option 1 is sometimes incorrectly referred to as "self-certify" or "self-certification". This is not an acceptable reference under ISO terms and definitions, for it can lead to confusion in the market. The user is responsible for making their own determination. Option 2 is often referred to as a customer or 2nd-party audit, which is an acceptable market term. Option 3 is an independent third-party process by an organization that is based on an engagement activity and delivered by specially trained practitioners. This option was based on an accounting procedure branded as the EnviroReady Report, which was created to help small- and medium-sized organizations. Its development was originally based on the Canadian Handbook for Accountants; it is now based on an international accounting standard. The fourth option, certification, is another independent third-party process, which has been widely implemented by all types of organizations. Certification is also known in some countries as registration. Service providers of certification or registration are accredited by national accreditation services such as UKAS in the UK.

ISO 14001 and EMAS

In 2010, the latest EMAS Regulation (EMAS III) entered into force; the scheme is now globally applicable, and includes key performance indicators and a range of further improvements. Currently, more than 4,500 organisations and approximately 7,800 sites are EMAS registered.

Complementarities and Differences

ISO 14001's environmental management system requirements are very similar to those of EMAS. Additional requirements for EMAS include:

- Stricter requirements on the measurement and evaluation of environmental performance against objectives and targets.
- Government supervision of the environmental verifiers.
- Strong employee involvement; EMAS organisations acknowledge that active employee involvement is a driving force and a prerequisite for continuous and successful environmental improvements.
- Environmental core indicators creating multi-annual comparability within and between organisations.
- Mandatory provision of information to the general public.
- Registration by a public authority.

ISO 14001 use in Supply Chains

There are many reasons that ISO 14001 should be potentially attractive to supply chain managers, including the use of the voluntary standard to guide the development of integrated systems, its requirement for supply chain members in industries such as automotive and aerospace, the potential of pollution prevention leading to reduced costs of production and higher profits, its alignment with the growing importance of corporate social responsibility, and the possibility that an ISO-registered system may provide firms with a unique environmental resource, capabilities, and benefits that lead to competitive advantage.

Emerging areas of research are starting to address the usc of this standard to show that ISO 14001 registration can be leveraged across the supply chain for competitive advantage. By looking at ISO 14001 registered firms, information from the study compared different amounts of integration and sustainability in the supply chain. Several research propositions and an empirical framework posit the impacts of ISO 14001 on supply chain design.

The propositions include:

- ISO registration leading to more proactive environmental management including process and performance measurement related to sustainability across a supply chain.
- That ISO-registered plants with formal environmental management systems will have higher levels of communication required between OEMs and Tier I suppliers.
- ISO-registered plants with direct relationships to other registered plants in their supply chain will have higher levels of waste reduction and cost efficiency than nonregistered plants.
- ISO-registered plants with direct relationships to other registered plants in the supply chain will have sustainable practices and projects with better ROI than nonregistered firms.
- ISO-registered plants with direct relationships to other registered plants will have higher levels of customer relationship management and will be positively associated with greater expansion opportunities and image than non-registered plants.

- ISO-registered plants with direct relationships to other registered plants will have fewer issues with employee health and reduced numbers of safety incidents than nonregistered plants.
- ISO-registered plants with a direct relationship to other registered plants will have a strong positive relationship between formal communication, training, monitoring/control systems, and firm performance.
- ISO-registered plants with a direct relationship to other registered plants will have higher levels of involvement and communication, which will be positively related to more internal and external integration with supply chain members.

List of ISO 14000 Series Standards

- ISO 14001 Environmental management systems - Requirements with guidance for use.
- ISO 14004 Environmental management systems - General guidelines on implementation.
- ISO 14006 Environmental management systems - Guidelines for incorporating ecodesign.
- ISO 14015 Environmental assessment of sites and organizations.
- ISO 14020 series (14020 to 14025) Environmental labels and declarations.
- ISO 14030 discusses post-production environmental assessment.
- ISO 14031 Environmental performance evaluation—Guidelines.
- ISO 14040 series (14040 to 14049), Life Cycle Assessment, LCA, discusses pre-production planning and environment goal setting.
- ISO 14046 sets guidelines and requirements for water footprint assessments of products, processes, and organizations. Includes only air and soil emissions that impact water quality in the assessment.

A water footprint assessment can assist in:

- Assessing the magnitude of potential environmental impacts related to water.
- Identifying opportunities to reduce water related potential environmental impacts associated with products at various stages in their life cycle as well as processes and organizations.
- Strategic risk management related to water.
- Facilitating water efficiency and optimization of water management at product, process and organizational levels.
- Informing decision-makers in industry, government or non-governmental organizations of their potential environmental impacts related to water (e.g. for the purpose of strategic planning, priority setting, product or process design or redesign, decisions about investment of resources).
- Providing consistent and reliable information, based on scientific evidence for reporting water footprint results.

A water footprint assessment alone is insufficient to be used to describe the overall potential environmental impacts of products, processes or organizations. The water footprint assessment according to this International Standard can be conducted and reported as a stand-alone assessment, where only impacts related to water are assessed, or as part of a life cycle assessment, where consideration is given to a comprehensive set of environmental impacts and not only impacts related to water. In this International Standard, the term "water footprint" is only used when it is the result of an impact assessment. The specific scope of the water footprint assessment is defined by the users of this International Standard in accordance with its requirements.

- ISO 14046 2014, Environmental Management- Water Footprint- Principles, Requirements, and Guidelines.
- ISO 14050 terms and definitions.
- ISO 14062 Integrating environmental aspects into product design and development (2002).
- ISO 14063 environmental communication guidelines and examples (2006).
- ISO 14064 measuring, quantifying, and reducing greenhouse gas emissions.
- ISO 19011 specifies one audit protocol for both 14000 and 9000 series standards together.

ISO 14064

The ISO 14064 standard (published in 2006) is part of the ISO 14000 series of International Standards for environmental management. The ISO 14064 standard provides governments, businesses, regions and other organisations with a complimentary set of tools for programs to quantify, monitor, report and verify greenhouse gas emissions. The ISO 14064 standard supports organisations to participate in both regulated and voluntary programs such as emissions trading schemes and public reporting using a globally recognised standard.

Structure of Standard

The Standard is published in three parts:

- ISO 14064-1:2006 specifies principles and requirements at the organization level for quantification and reporting of greenhouse gas (GHG) emissions and removals. It includes requirements for the design, development, management, reporting and verification of an organization's GHG inventory.
- ISO 14064-2:2006 specifies principles and requirements and provides guidance at the project level for quantification, monitoring and reporting of activities intended to cause greenhouse gas (GHG) emission reductions or removal enhancements. It includes requirements for planning a GHG project, identifying and selecting GHG sources, sinks and reservoirs relevant to the project and baseline scenario, monitoring, quantifying, documenting and reporting GHG project performance and managing data quality.
- ISO 14064-3:2006 specifies principles and requirements and provides guidance for those

conducting or managing the validation and/or verification of greenhouse gas (GHG) assertions. It can be applied to organizational or GHG project quantification, including GHG quantification, monitoring and reporting carried out in accordance with ISO 14064-1 or ISO 14064-2.

ISO 14064-3:2006 specifies requirements for selecting GHG validators/verifiers, establishing the level of assurance, objectives, criteria and scope, determining the validation/verification approach, assessing GHG data, information, information systems and controls, evaluating GHG assertions and preparing validation/verification statements.

The principles behind ISO 14064 have been used in national calculation methodologies such as the UK's Carbon Trust Standard.

Ems Standard: Iso 14000 Series

Companies throughout the world have begun to understand the requirements of an environmental management system (EMS) as well as the benefits such a system can provide. However, given the nature of company dynamics, changes to business priorities, personnel and work patterns, it is imperative to introduce a proper management system which maintains the focus on environmental improvement through the changing landscape of corporate activity. It is against this background that ISO 14000 Series are being developed and implemented.

Evolution

The International Organisation for Standardisation (ISO) was formed in 1947 and has since become the premier international standards organisation. Since its inception, ISO's mission has been to promote worldwide standardisation in order to facilitate international commerce. The organisation began by developing international agreements and in 1951 published its first standard, Standard Reference Temperature for Industrial Length Measurement. Since that time, ISO has developed more than 9,000 standards for a variety of subjects ranging from screw threads and fasteners to high-tech clean rooms. ISO also developed the widely used quality management system (QMS) standards, i.e., the ISO 9000 series. The ISO standards are published for voluntary acceptance, but they are often incorporated into national standards of individual countries.

The membership of ISO includes over 100 countries. Each Member Country is represented by one standards organisation. For example, the American National Standards Institute (ANSI) represent the U.S and Great Britain by the British Standards Institute (BSI).

ISO considers the following three key principles in developing international standards:

(i) Consensus: The views of all interested parties are taken into account, including manufacturers, vendors, consumer groups, testing laboratories, governments, engineering professionals and research organisations.

(ii) Industry-wide applicability: The goal is to draft standards that satisfy industries and customers worldwide (ISO has no authority to impose its standards on any government or organisation).

(iii)Voluntary nature: All of the international standards developed are voluntary. Thus, their acceptance by industry is market-driven and based on voluntary involvement of all interests in the marketplace.

A review of historical developments reveals that standardisation on a worldwide basis was accelerated with quality and occurred generally independent of environmental management. Attempts to standardise quality requirements were made by many organisations. However, it wasn't until the 1990s that a significant level of agreement was reached.

In 1990, Business Charter for Sustainable Development (BCSD), an organisation of fifty business leaders with interest in environment and development issues, was created for environment protection with the premise that economic development can take place only in a healthy environment. Partly in response to the proliferation of differing environmental standards, such as EMAS worldwide, the ISO formed a Strategic Action Group on the Environment (SAGE) in 1991.The purpose of SAGE was to investigate the usefulness of drafting international standards for environmental management. SAGE focused its attention on the following three areas:

(i) Promoting a common worldwide approach to environmental management in business and industry.

(ii) Increasing the ability of incentives for organisations to measure and attain improvements in environmental performance.

(iii)Facilitating world trade and removing potential environmental trade barriers.

As a result of the findings of SAGE, the ISO formed Technical Committee 207 (TC 207) in 1993. TC 207 became responsible for drafting the ISO 14000 series of standards. At early meetings of TC 207, more than thirty countries and 200 representatives expressed a desire to develop new EMS standards. TC 207 itself had members representing some sixty-nine countries and was divided into six sub-committees. These members included representatives from various industries, standards organisations, governments, environmental organisations, and other interest groups.

For some, the motivation for the development of these standards was due to the fear that the increased number of inconsistent national and regional EMS standards would create trade barriers. There was also the concern that the EMAS programme already in practice in Europe would influence the ISO 14000 standards to make them comparable to EMAS. This concern was in part fueled by the fact that the ISO 14000 standards were being developed in conjunction with CEN, i.e., the European standards-setting body. In fact, the ISO set the standards drafting timetable at 30 months, in part because CEN had agreed to accept ISO 14000, if they were finalised quickly enough. There were also fears that the standards were a reflection of European not American technology. Nevertheless, the standards were developed with each participating member having equal say in the process.

A sub-committee of TC 207 prepared the draft EMS Standard, based on BS 7750. It was presented at the Earth Summit in Rio de Janeiro, and the draft of the standard was agreed in Oslo (June 1995).

Principles and Structure

The aim of the ISO 14000 series of standards is to help organisations implement and improve their EMS.

Some of the principles governing ISO 14000 series are (Tech Monitor, Vol. 12. No. 5 Sept-Oct 1995):

(i) Understand all activities and processes being undertaken by the organisation.

(ii) Identify potential aspects associated with the activities of the unit at all stages and determine their impact on the environment.

(iii) Determine processes/procedures/operation steps that can be controlled to eliminate or minimise the likelihood of an occurrence of the adverse impact.

(iv) Identify the regulatory requirements relating to them and establish target level and tolerances, which must be met to ensure that operations affecting the environment are under control.

(v) Establish a monitoring mechanism to ensure control of these aspects.

(vi) Establish corrective actions to be taken when monitoring indicates that a particular aspect is not under control.

(vii) Establish a system of emergency preparation and for meeting such exigencies.

(viii) Establish procedures for verification to confirm that the environmental management system is working effectively in compliance with regulations and recording continuous improvement.

(ix) Establish documentation concerning all procedures and records appropriate to these principles and their application.

Given the importance of the stakes involved and the generic nature of the international requirements covered, high expectations have been placed on the ISO 14000 series standards. The standards can be classified into the following two categories based on their focus:

(i) Organisation or process standards: These include environmental management system (EMS), environmental auditing (EA) and environmental performance evaluation (EPE).

(ii) Product-oriented standards: These include life-cycle assessment (LCA), environmental labelling (EL) and environmental aspects in product standards (EAPS).

The ISO 14000 series standards are of two types, and these are: (i) Normative standards: These indicate the requirements that must be met and can be audited for certification. (ii) Informative standards: These provide guidance and the requirements need not be audited for certification.

In the 14000 series of standards, ISO 14001, (i.e., Environmental Management Systems-specification with guidance for use) is the only normative standard and all other standards are informative standards intended to support the implementation of EMS.

As regards the structure of the Series, ISO 14001, i.e., the requirements for EMS, forms the nucleus, and in the first orbit is ISO 14004, i.e., the guideline standard. The other standards in the series are supporting systems.

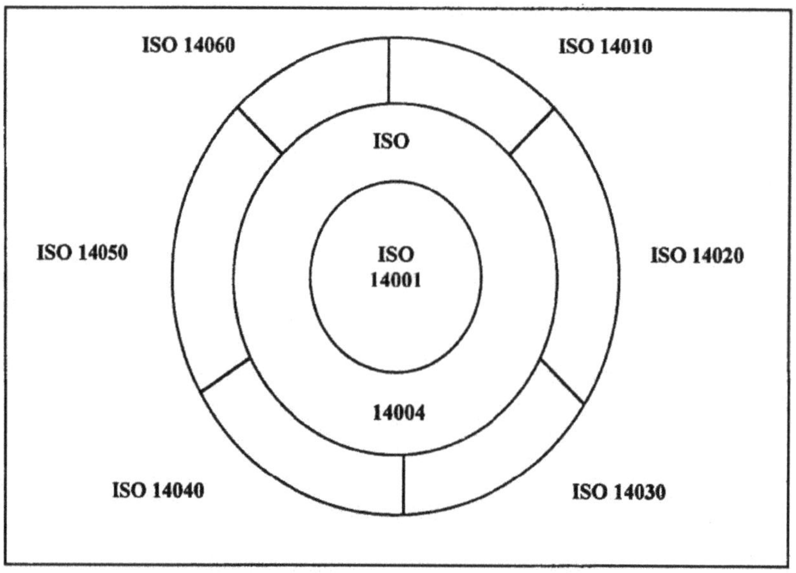

Structure of ISO Series and Inter-relationships

Supporting Systems

The supporting systems of ISO 14000 consist of ISO 14010, ISO 14020, ISO 14030, ISO 14040, ISO 14050 and ISO 14060.

Consider the table, which gives a list of 150 14000 Series standards with their respective publication dates.

ISO 14000 Series Standards:

ISO No.	Title of International Standard/Guideline/Technical Report
ISO 14001	Environmental management
ISO No.	Title of International Standard/Guideline/Technical Report
	systems – Specification with guidance for use.
ISO 14004	Environmental management systems – General guidelines on principles, systems and supporting techniques.
ISO/AWI 14004	Revision of ISO 14004:1996.
ISO 14010	Guidelines for environmental auditing – General principles.
ISO 14011	Guidelines for environmental auditing - Audit procedures – Auditing of environmental management systems.

ISO 14012	Guidelines for environmental auditing – Qualification criteria for environmental auditors.
ISO 14015	Environmental management – Environmental assessment of sites and organisations (EASO).
ISO/DIS 19011	Guidelines for quality and/or environmental management systems auditing.
ISO 14020 2nd Edition	Environmental labels and declarations – General principles.
ISO 14020: 1998/DAM1	Draft amendment 1 to ISO 14020:1998.
ISO 14021	Environmental labels and declarations - Self-declared environmental claims (Type II environmental labelling).
ISO 14024	Environmental labels and declarations - Type I environmental labelling – Principles and procedures.
ISO/TR 14025	Environmental labels and declarations – Type III environmental declarations.
ISO 14031	Environmental management – Environmental performance evaluation – Guidelines.
ISO/TR 14032	Environmental management – Examples of environmental performance evaluation.
ISO No.	Title of International Standard/Guideline/Technical Report
ISO 14040	Environmental management – Life cycle assessment – Principles and framework.
ISO 14041	Environmental management – Life cycle assessment – Goal and scope definition and inventory analysis.
ISO 14042	Environmental management – Life cycle assessment – Life cycle impact assessment.
ISO 14043	Environmental management – Life cycle assessment – Life cycle interpretation.
ISO/WD TR 14047	Environmental management – Life cycle assessment – Examples of application of ISO 14042.
ISO/CD TR 14048	Environmental management – Life cycle assessment – Life cycle assessment data documentation format.

ISO/TR 14049	Environmental management – Life cycle assessment – Examples of application of ISO 14041 to goal and scope definition and inventory analysis.
ISO 14050	Environmental management – Vocabulary.
ISO 14050: 1998/DAM1	Draft amendment 1 to ISO 14050:1998.
ISO/TR 14061	Information to assist forestry organisations in the use of Environmental Management System standards ISO 14001 and ISO 14004.
ISO/AWI 14062	Guidelines for integrating environmental aspects into product development.
ISO Guide 64	Guide for the inclusion of environmental aspects in product standards.

Environmental Auditing (ISO 14010, ISO 14011 and ISO 14012)

An environmental audit is defined as the systematic documented verification process of objectively obtaining and evaluating audit evidence to determine whether specified environmental activities, processes, conditions, management systems, or information about these matters conform with audit criteria, and communicating the results of this process to the clients. The guiding principles of environmental audits include:

- basing the audit on defined objectives and drawing inferences based on analysis, interpretation and documentation of appropriate information;
- Utilising an audit team that is independent of the activities they audit and utilising an auditor who meets the specifics of qualification criteria;
- Exercising of due professional care by the auditor to maintain confidentiality and adequate quality assurance;
- Using appropriate procedures for an objective audit;
- Developing audit criteria, evidence and findings;
- Ensuring that the process provides a desired level of confidence in the reliability of the audit findings and conclusions;
- Providing an adequate report of findings.

Note that while ISO 14010 provides guidance on general principles for conducting environmental audits, ISO 14011 elaborates on the framework for the conduct of audit in order to ascertain if the organisation is doing what it says it will do and whether the EMS conforms to ISO 14001. ISO 14012 provides guidance on auditor qualification criteria, including education, training, work experience, personal attributes and skills, maintenance of competence and due diligence.

Environmental Labelling Standards (ISO 14020)

ISO 14020 is based on a voluntary environmental labelling standard that requires a third party verification and is designed to reduce burdens arising from diverse multiple labelling. Existing eco-labelling schemes, based on government initiatives, aim at influencing customer decisions to select environmentally friendly products, but they lack the application of uniform criteria. These are, therefore, difficult to comply with and are potential trade barriers. The coverage of ISO 14020 is broad and includes goods or services for consumer, commercial and industrial purposes.

The guiding principles and practices for third-party environmental labelling programmes include the following:

- Standard and criteria applicable for environmental labels must be developed through a consensus process, and the programme must be voluntary.
- A product can be considered for environmental regulations of the country in which it is manufactured and marketed.
- Environmental labelling programmers should be selective and should distinguish leading product alternatives.
- The product criteria developed by the certification agency should be periodically reviewed to account for new developments and technologies.
- The process should be transparent with regard to criteria, certification and award procedures.
- The criteria should be revised periodically and clearly demonstrate that the funding sources for the programme do not create conflicts of interest or undue influence.
- The labelling programme should use scientific and reproducible methodologies to assess the environmental impacts of products.
- Labelling programmes should be accessible, objective and affordable.

The requirements for awarding a label are divided into general rules that apply to all products and applications, and specific product criteria that set requirements for each product category. The specific product criteria are the only criteria that may be considered as a basis for awarding the label. The certification agency awards the label when satisfied that the applicant has complied with the specific product criteria for the category. It maintains a publicly available list of products currently licensed to carry the label.

After the label has been awarded, it is the responsibility of the certification agency to take all necessary steps to ensure ongoing compliance with the product criteria. The certification agency will require the licensee to take corrective action, if monitoring indicates that compliance is not being maintained.

Environmental Performance Standards (ISO 14030)

The environmental performance evaluation (EPE) is an important ongoing internal management process. It uses environmental indicators to compare an organisation's past and present environmental performance with it's environmental objectives, targets or other intended levels of environmental performance. This process helps management to measure, analyse, assess, report and communicate an organisation's environmental performance over time and to determine necessary actions.

The environmental performance evaluation standard is based on Edward Deming's PDCA cycle, i.e., Plan (P), Do (D), Check (C) and Act (A). Planning a EPE involves considering management priorities and selecting environmental indicators before evaluating environmental performance (D) and reviewing and improving EPE (C). The EPE process involves actions taken (A) to improve the systems, operations, processes and environmental performance.

EPE standards provide guidelines on identification and selection of environmental indicators. There are two types of environmental indicators, and these are:

(i) Environmental performance indicators (EPI): These include the following:

- The people, practices and producers at all levels of the organisation;
- The design, operation and maintenance of, supply to, and delivery from, an organisation's facilities and equipment;
- The materials, energy, products, services, waste and emissions related to the organisation's operations and activities.

(ii) Environmental condition indicators (ECI): These are used to describe the conditions of the environment in relation to the organisation.

Life Cycle Assessment (ISO 14040)

ISO 14040 provides guidelines for incorporating life cycle assessment (LCA) into environmental management programmes (recall that in Unit 6 we discussed LCA in detail). An LCA is defined as a compilation and evaluation of the inputs, outputs and the potential environmental impacts of a product system throughout its life cycle. It is, thus, a systematic set of procedures for compiling and examining the inputs and outputs of materials and energy and the associated environmental impacts directly attributable to the functioning of a product or service system throughout its life cycle from the acquisition of raw materials through final disposal. An LCA is done in order to get the whole picture of the environmental impacts throughout the lifetime of products and services. In other words, an LCA provides significantly more information than does evaluating the impact from the manufacturing process alone. It also provides a systematic way to evaluate the costs and benefits associated with products or service changes at various points in the life cycle.

ISO 14040 covers the following three phases of LCA:

- Establishing the goals and the scope of the assessment;
- Conducting inventory analysis;
- Conducting impact and improvement assessment.

The goals of the LCA study should include the reasons for carrying out the study, the intended applications, the intended audience, the initial data, quality objectives and the type of critical review that will be conducted for the LCA. The scope should include background information for the product or service being evaluated, boundaries of the study, method of impact assessment, data requirements, assumptions and limitations of the study.

Environmental Terms and Definitions (ISO 14050)

The successful operation of the environmental management system requires a correct understanding of the meaning of terms used by all stakeholders in a similar manner. Put differently, communication is important in implementation and in the operation of the environmental management system. This communication will be most effective, if there is a common understanding of the terms used. Many environmental terms and definitions are the result of a recently developed process. The gradual evolution of these environmental concepts invariably means that environmental terminology will continue to develop.

EMS Specification Standards: ISO 14001

This standard is the core of the ISO 14000 series. ISO 14001 presents the elements of a EMS that organisations are required to conform with, if they seek third-party certification. As mentioned previously, ISO 14001 is the only standard in the ISO 14000 series intended to be used for third-party certification. All the other standards in the series are intended as guidance only. In development of your EMS, ISO 14001 will be your most important document – you will refer to it frequently.

The standard consists of the EMS specification in 17 general requirements (referred to as clauses) and three annexes.

Clauses

Because the standard has been written to be applicable to all types of organisations and to accommodate diverse geographical, cultural and social conditions, the requirements or clauses, are generalised and are neither specific nor prescriptive. That is to say, they are descriptive outlining the desired outcome of the system, but do not prescribe specific approaches that an organisation must implement to get there.

Annexes

Annex A provides additional information on the requirements of ISO 14001 and is intended to help avoid any misinterpretation of the actual specification. It is expected that in future revisions to ISO 14001, the information in Annex A may be either incorporated into the clauses or moved into ISO 14004. Annex B identifies links and broad technical correspondences between ISO 14001 and ISO 9001 in a tabular form. The objective of this annex is to demonstrate how a EMS could be combined with a quality management system. Annex C simply presents a bibliography of documents from both the ISO 9000 and ISO 14000 series.

Note that these annexes are solely for information and not for third-party certification. ISO 14001 is a management system standard. It is not a performance or product standard, although the framers developed it with the idea that the result of its implementation should give better industry environmental performance. The standard represents a shift towards holistic proactive management and total employee involvement. ISO 14001 urges employees to define their environmental roles from the bottom up and requires the support of senior management. It is a comprehensive framework that contains core elements for managing an organisation's processes and activities to identify significant environmental aspects the organisation can control, and over which it can be expected to have an influence. Any organisation or facility or organisation of any size anywhere in the world can use the standard.

An EMS conforming to ISO 14001 contains the following elements:

- An environmental policy supported by senior management.
- Identification of environmental aspects and significant impacts.
- Identification of legal and other requirements.
- Environmental goals, objectives and targets that support the policy.
- An environmental management programme.
- Definition of roles, responsibilities and authorities.
- Training and awareness procedures.
- Process for communication of the EMS to all interested parties.
- Document and operational control procedures.
- Procedures for emergency response.
- Procedures for monitoring and measuring operations that can have a significant impact on the environment.
- Procedures to correct non-conformance.
- Record management procedures.
- A programme for auditing and corrective action.
- Procedures for management review.

Note that while ISO 14001 is more likely to be used in manufacturing or processing industries, it can also be applied to services such as construction, architecture, health care and engineering. This standard essentially requires an organisation to state what it does in environmental management and do what it states. ISO 14001 is neither a product standard nor an environmental performance standard. It does not require an organisation to establish or disclose performance or audit results and does not require certification.

Let us now discuss the guidelines for planning and implementation of ISO 14001.

Implementation oF Ems Conforming To Iso 14001

The key to a successful ISO 14001 EMS is having documented procedures that are implemented and maintained in such a way that successful achievement of environmental goals is commensurate with the nature and scale of the activities.

There are three components to a EMS and these are:

(i) A written programme: The written programme requires an organisation to be committed to producing a quality product with the lowest possible environmental impact and sets forth the procedures to be followed to achieve this goal.

(ii) Education and training: The implementation will not be effective, unless all employees have access to and understand the EMS, and this is achieved through education and training.

(iii) Knowledge of relevant local and central environmental regulations: The EMS must incorporate the applicable environmental legal requirements.

In addition, the EMS must include appropriate monitoring and review to ensure effective functioning of the EMS and to identify and implement corrective measures in a timely manner.

The various steps involved in the establishment of a EMS are as follows:

(i) Obtain management commitment: Senior management can demonstrate its commitment by playing an active, visible role in the EMS implementation process, providing funding and allocating resources, and promoting employee awareness and motivation. A EMS should be viewed as a tool to achieve continuous environmental improvement, rather than daily fire-fighting just to meet the mandatory requirements.

(ii) Review current environmental programmes: Once the management commitment is secured, the next step is to conduct and document a preliminary review of the organisation's current environmental programmes and management systems. The process includes reviewing existing environmental management policies, operating procedures and training programmes as well as the methods for identifying regulatory requirements.

(iii) Conduct gap analysis: Organisations need to evaluate as to what extent the existing procedures conform to the requirements of ISO 14001. Such a gap analysis identifies the actions necessary to build a EMS. To do this, organisations would need to compare their existing procedures to the following main stages of ISO 14001:

- Environmental policy.
- Planning.
- Implementation and operation.
- Checking and corrective action.
- Management review.
- Continuous improvement.

(iv) Identify environmental aspects: All parts of the planning phase contribute to identifying the significant environmental aspects of the operations. An environmental aspect refers to any element of an organisation's activities, products and services that can interact with the environment. ISO 14001 also refers to impacts. Impacts are the actual or potential changes to the environment resulting from any of the environmental aspects. Thus, the relationship between aspects and impacts is one of cause and effect. One of the best ways to identify the environmental aspects of the operation is by developing a process map or flowchart. This exercise involves mapping every step of the process and the inputs and outputs associated with each step. Developing a process map is best accomplished as a team effort, and therefore, representatives of a variety of departments and at diverse levels should be included to produce the most accurate description of the inputs and outputs. Note that an organisation's EMS is a team effort.

(v) Identify significant aspects: An organisation can utilise the expertise of staff to determine the significant environmental aspects of its operations. This will help in focusing on those aspects over which the organisation has some control or influence. Determining which aspects are significant is based on the judgement of severity of the impact caused by the aspect and the frequency of its occurrence, and therefore, each aspect needs to be scrutinised. It requires many iterations, before the team arrives at a final list of significant environmental aspects.

(vi) Decide the environmental policy: Having identified the significant environmental aspects and established the means and norms to improve environmental performance in the form of objectives and targets, organisations need to write their broad intentions concerning the environment. This will form or evolve into an appropriate policy, meeting the requirements of the standards. This document must include the organisation's commitments towards complying with legal requirements, pollution prevention and continuous improvement. Note that policy framing may not be a onetime affair. It is quite likely that organisations are required to revisit the objectives formulated and targets identified commensurating with the outcomes of the initial implementation process. This iterative process helps in firming up the policy statements.

(vii) Identify environmental management programmes: An environmental management programme (EMP) may be considered a project or a plan of action to achieve objectives and targets. At this stage, the methodology of realising the objectives, the steps involved, the persons responsible for implementing these steps, the time frame of implementation, the means required, etc., should be noted.

(viii) Identify training needs and personnel whose work affects the environment: It is essential to identify the training needs of the personnel by comparing the present competence (based on the educational qualifications, past work experience, etc.) with the required competence. Organisations are expected to have training needs, training plans, training modules and training records, and must check the effectiveness of the training imparted. The training must make the personnel aware of the importance of conformance with environmental policy, environmental impacts of their work activity and the consequences of departing from the specified operating procedures.

(ix) Establish communication system: Proper communication channels are imperative to impress on the employees, as well as external interested parties, the reasons for adopting EMS.

(x) Create proper documentation: The EMS documentation describes the interaction among the core elements of the system and provides the key information necessary to understand the environmental management systems in place at the organisation. This is of immense use to the new employees and third-party auditors.

(xi) Conduct EMS audits: Organisations are required to conduct EMS audits to verify the EMS implementation. But, EMS audits conducted merely to meet the requirements of the ISO standard will not be much use in the long term. For sustained benefits, organisations must invest in EMS audit programmes by giving employees proper training, providing appropriate time for audit, etc. Importantly, senior management should demonstrate its commitment to audits and provide visible support to the auditors. This will encourage the auditors to point out the weaknesses in the system without any hesitation. As inept audits – as a result of doctored auditing – are not uncommon, the support of senior management is very critical in carrying out and documenting EMS audits.

(xii) Conduct management review: Management must periodically review the EMS to evaluate its suitability and effectiveness.

(xiii) Perform a final gap analysis: Final gap analyses help organisations determine whether they are ready for a registration audit. In other words, the results of the final gap analyses help organisations decide whether or not to apply for certification by outside auditors.

Registration

Organisations are registered by outside auditors – also known as registrars – meeting the ISO 14001 standard. The registration can be for a specific site, several sites or the entire organisation. The registration process involves the following steps:

(i) Establishing the scope of the registration. This can be done internally or with the help of outside auditors.

(ii) Applying for registration. This is a written application to the registration body.

(iii) Reviewing documents about the organisation's EMS and submitting them to the registration body. Outside auditors perform full assessment.

(iv) Getting the grant of registration by the certification body based on the auditor's findings.

(v) Receiving either approval (conditional or provisional) or disapproval.

(vi) Receiving a certificate as a proof of registration and getting listed in a certification register.

(vii) Maintaining the registration through ongoing activities including monitoring and measuring the EMS, investigating and handling non-conformances, implementing corrective and preventive actions, maintaining environmental records and establishing and maintaining the ISO 14000 audit programme.

The time required to implement ISO 14001 EMS depends on the current status. That is to say, if an organisation's environmental programmes are effective and efficient and they are willing to commit resources, the EMS development will be fast. If all the factors required for the EMS development are in place, it is possible to complete the entire process till certification in eight to ten months. (Note that it may take the equivalent of one fourth of an employee's time for preliminary planning and exploring issues. The final four to five months will be required for checking and improving the established system. The longest and most timeconsuming activity in the entire process will be training.)

Before you read any further, you must note that:

- ISO does not assess or audit environmental management systems to confirm that they conform to ISO 14001.
- ISO does not issue ISO 14000 certificates.
- ISO does not carry out ISO 14000 certificates (these are issued by certification (registration) bodies independently of ISO).
- The ISO logo is a registered trademark and ISO does not authorise its use in connection with ISO 14000 certification (registration).
- ISO 14001 is not a label signifying a green or environmentally friendly product. (No product label, advertisement or other promotional material should give the impression that a product is "ISO 14001-certified" or "ISO 14001-registered". Advertisements, which imply a quality ranking, i.e., ISO 9001, ISO 9002 and ISO 9003, are misleading.)

Benefits of Implementing Iso 14001: An Indian Scenario

Confederation of Indian Industries (CII) is the nodal agency for ISO 14000 in India. By 2001, over 350 organisations in India have established EMS conforming to ISO 14001.

Sector-Wise Distribution Of 14001 Certified Enterprises:

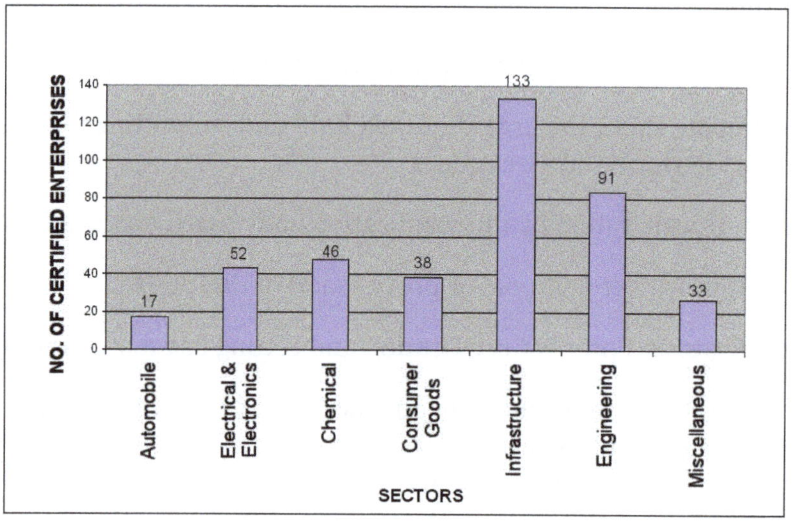

Based on the reports of CII, we will list below the benefits perceived by the organisations after implementing ISO 14001 environmental management system.

- Environmental benefits: These include identification of new opportunities for environmental improvements and improvement in resource conservation and waste minimisation.
- Cost containment and cost savings: These include identification of new opportunities for cost savings and areas of improvement.
- Regulatory compliance: This includes risk reduction and a lower rate of regulatory enforcement.

- Competitiveness: This includes an increase in market share and a decrease in cost of production.
- International trade: This includes an increase in exportrelated business opportunities.
- Greening the suppliers: These include the opportunities for greening the supplier chain. (Note that this has been a major thrust area for those firms who have a wide range of suppliers. At present, most firms generally do not include environmental issues in their purchase contracts. Firms have to impart education/training and transfer of know-how, before actually asking the supply chain to go green. Before long, the suppliers' environmental performance will become important for maintenance of quality and compliance of environmental standards.)
- Other benefits: These include achievement of commitment to good environmental performance at all levels, safe and healthy working conditions as well as better housekeeping.

Some specific cases of firms that have benefited from the establishment of EMS conforming to ISO14001 are given below:

- The case of an international electrical and electronic goods giant. It has a very comprehensive and ambitious EMS programme for more than 200 of its factories worldwide supported by the parent body. Other than ISO 14001, it also conforms to a headquarter-driven quality programme. The environmental aspects recognised by the organisation were development and production of different types of capacitors; manufacture of lamps along with their components and manufacture of audio products, PCBs and electronic and electromagnetic ballast. The most important motivation factors were directions from the headquarters; enhancing the organisation's image; resource conservation benefits and ensuring better compliance with legal requirements. After the implementation of EMS, the organisation identified some intangible benefits, which had not been included in its original goal. These benefits were better interaction among suppliers, employees and authorities; improved waste management practices; improved employee awareness and enhanced staff morale; better working conditions and greater domestic share. The major tasks the organisation accomplished after implementation of EMS included 50% reduction in water consumption; 12 - 25% energy saving depending upon the plant; reduction in packaging mass and reductions in emissions by 95%.
- The case of an organisation was certified to Responsible Care before it was certified to the ISO 14001 protocol. The organisation achieved cost savings from pollution prevention, energy conservation and waste minimisation. The savings included recovery of various products (Rs. 5,55,000), reduction in water, fuel, electricity consumption (Rs. 17,10,000), reuse (Rs. 7,00,000), and others. Note that though the organisation followed the Responsible Care norms, it was better able to recognise potential areas for cost savings through the adoption of ISO 14001.
- The case of a mining organisation that maintains the heavy hydraulic machinery/vehicles used for its excavation work in a workshop situated on top of a hill. As part of the maintenance, oil changes were made and oil was carelessly being thrown on the land, thus contaminating it. Through the EMS, an organisation-wide oil balance study was initiated and

it was found that there was a large gap between the oil received and the waste oil generated by the workshop. On a closer study, it was found that the gap accounted for the oil spill on the land. Consequently, the operating practice was changed and trays were installed to collect the waste oil. From the trays, the waste oil was transferred to drums and were sold. This sale resulted in a revenue flow of Rs. 36 lakhs in the first year of the organisation's EMS implementation.

- The case of a paint-manufacturing firm. At this time, there was no monitoring of the amount of water consumed, and different estimates of water consumption ranging between 250 to 400m3 /day were available. Through the EMS, the organisation actually found that it was 490 m3 /day. It decided to make a significant resource saving by reducing water wastage. After plugging the water leakage points and controlling the overflow in the cooling tower, water consumption was reduced to 210 m 3 /day, i.e., a saving of 270m3 /day.
- The case of an electrical cable manufacturing organisation. This organisation makes aluminum core cable with an XLPE (cross-linked polyethylene) coating, which is a black hard plastic material. In the process of making this aluminum core, there was the mechanical drawing of the wire from a metal block. This operation generates fine aluminum dust and was inhaled by the workers in the plant (as this metal dust was not contained), which caused respiratory problems and further led to the absenteeism of the workers. Through the EMS, the organisation decided to contain this waste generation by encasing the metal drawing operation. The aluminum dust generated was properly collected and sold to a local paint manufacturer who used it to manufacture silver paint.
- The case of a leading cloth mill. This organisation is the largest denim-manufacturing firm in India. It was the first denim manufacturer in India to gain the ecologically optmised fabric (EOF) trademark an eco-tex certification. It exports nearly 75 – 80% of its total denim production. The most important motivating factors for the firm were enhancing the organisation's image; augmenting management systems culture from quality to other areas of social concern and ensuring legal compliance and achieving higher environmental performance standards. During the implementation phase, the organisation's main focus was on resource conservation. The benefits that the firm realised after the implementation of ISO 14001 include graduation from ad hoc arrangements to using systems approach for environmental management that is consistent with the overall operations of the unit, incorporation of environmental parameters in the corporate decision-making process and enhancement of the organisation's image and greater worker motivation.

OHSAS 18001

OHSAS 18001, Occupational Health and Safety Assessment Series, (officially BS OHSAS 18001) is an internationally applied British Standard for occupational health and safety management systems. It exists to help all kinds of organizations put in place demonstrably sound occupational health and safety performance. It is a widely recognized and popular occupational health and safety management system.

This Certificate is to verify that:

Borj Pardis

Have been audited by:

the audit people

And has been assessed for conformance to the requirements of:

OHSAS 18001:2007

Occupational Health and Safety Management System.

Scope of Registration

General Contracting, Engineering, Procurement & Constructing of Various Projects

Registration Number

4020959

Signed on Behalf of The Audit People Ltd
Certification Manager: Elaine Hanaghan

Initial Certification Date:	**12th April 2012**
Certificate Issue Date:	**13th April 2012**
Certificate Expiry Date:	**11th April 2015**
Issue No.	**01**

Certificate of Registration

Certificate of conformity to OHSAS 18001:2007

Organizations worldwide recognize the need to control and improve health and safety performance and do so with occupational health and safety management systems (OHSMS). However before 1999 there was an increase of national standards and proprietary certification schemes to choose from. This caused confusion and fragmentation in the market and undermined the credibility of individual schemes.

Development

Recognizing this deficit, an international collaboration called the Occupational Health and Safety Assessment Series (OHSAS) Project Group was formed to create a single unified approach. The Group comprised representatives from national standards bodies, academic bodies, accreditation bodies, certification bodies and occupational safety and health institutions, with the UK's national standards body, BSI Group, providing the secretariat.

Drawing on the best of existing standards and schemes, the OHSAS Project Group published the OHSAS 18000 Series in 1999. The Series consisted of two specifications: 18001 provided requirements for an OHS management system and 18002 gave implementation guidelines. As of 2005, around 16,000 organizations in more than 80 countries were using the OHSAS 18001 specification. By 2009 more than 54,000 certificates had been issued in 116 countries to OHSAS or equivalent OHSMS standards.

Adoption as British Standard

The OHSAS 18001 specification was updated in July 2007. Among other changes, the new specification was more closely aligned with the structures of ISO 9000 and ISO 14000 so that organizations could more easily adopt OHSAS 18001 alongside existing management systems. Additionally, the "health" component of "health and safety" was given greater emphasis.

Later, the BSI Group decided to adopt OHSAS 18001 as a British standard. BSI Group subsequently adopted the updated 18002 guidance specification for publication as BS OHSAS 18002 in 2008.

How They Work

Its supporters claim that an occupational health and safety management system (OHSMS) promotes a safe and healthy working environment by providing a framework that helps organizations to:

- Identify and control health and safety risks.
- Reduce the potential for accidents.
- Aid legal compliance.
- Improve overall performance.

The OHSAS 18000 standards provide organizations with the elements of an effective safety management system which can be integrated with other management systems and help organizations achieve better occupational health and safety performance and economic objectives.

BS OHSAS 18001 specifies requirements for an OH&S management system to help an organization develop and implement a policy and objectives, which take into account legal requirements and information about OH&S risks. It applies to all types and sizes of organizations and accommodates diverse geographical, cultural and social conditions.

BS OHSAS 18002 provides guidance for establishing, implementing or improving a management system which is based on OHSAS 18001 and demonstrating successful implementation of OHSAS 18001.

OHSAS 18001 can be aligned with existing ISO 9001 and ISO 14001 management systems. Historically many organizations start with the quality management system ISO 9001, then add the environment management requirements from ISO 14001. Many organizations now look at implementing all three standards at once which can minimize costs and disruption. The standards can be integrated using a standard such as BSI's PAS 99.

Disambiguation

The OHSAS 18000 standards were written and published wholly outside of the International Organization for Standardization (ISO) framework. To avoid confusion, ISO 18000 does exist – but it is a radio-frequency identification standard.

ISO 45001

In October 2013, the International Organization for Standardization approved the project proposal to develop the ISO 45001 standard, an ISO analogue to the OHSAS 18000 standards. This is stated to be taking OHSAS 18001 into account along with other international standards.

OHSAS 18001 and its Comparison with ISO 14001 and ISO 9001

The (OHSAS) specification gives requirements for an occupational health and safety (OH&S) management system, to enable an organisation to control its OH&S risks and improve its performance.

It does not state specific OH&S performance criteria, nor does it give detailed specifications for the design of a management system. OHSAS 18001 has been developed to be compatible with the ISO 9001 (Quality) and ISO 14001 (Environmental) management systems standards, in order to facilitate the integration of quality, environmental and occupational health and safety management systems by organizations.

Scope

This Occupational Health and Safety Assessment Series (OHSAS) specification gives requirements for an occupational health and safety (OH&S) management system, to enable organizations to control its OH&S risks and improve performance. It does not state specific OH&S performance criteria, nor does it give detailed specifications for the design of a management system.

This OHSAS specification is applicable to any organization that wishes to:

a) Establish an OH&S management system to eliminate or minimize risk to employees and other interested parties who may be exposed to OH&S risks associated with its activities;

b) Implement, maintain and continually improve an OH&S management system;

c) Assure itself of its conformance with its stated OH&S policy;

d) Demonstrate such conformance to others;

e) Seek certification/ registration of its OH&S management system by an external organization; or

f) Make a self-determination and declaration of conformance with this OHSAS specification.

This OHSAS specification is intended to address occupational health and safety rather than product and service safety.

Objectives

The organization shall establish and maintain documented occupational health and safety objectives, at each relevant function and level with the organization. This needs to be quantified wherever practicable.

When establishing and reviewing its objectives, an organization shall consider its legal and other requirements, its OH&S hazards and risks, its technological options, its financial, operational, and business requirements, and the views of interest parties. The objectives shall be consistent with the OH&S policy, including the commitment to continual improvement.

A main driver for this was to try to remove confusion in the workplace from the proliferation of certifiable OH&S. specifications. This is based on the concerted effort from a number of the world's leading national standards bodies, certification bodies, specialist consultancies and published documents:

- BS8800:1996 Guide to occupational health and safety management systems.
- Technical Report NPR 5001: 1997 Guide to an occupational health and safety management system.

- SGS & ISMOL ISA 2000:1997 Requirements for Safety and Health Management Systems.
- BVQI SafetyCert: Occupational Safety and Health Management Standard.
- DNV Standard for Certification of Occupational Health and Safety Management Systems(OHSMS):1997.
- Draft NSAI SR 320 Recommendation for an Occupational Health and Safety (OH and S) Management System.
- Draft AS/NZ 4801 Occupational health and safety management systems Specification with guidance for use.
- Draft BSI PAS 088 Occupational health and safety management systems.
- UNE 81900 series of pre-standards on the Prevention of occupational risks.
- Draft LRQA SMS 8800 Health & safety management systems assessment criteria.

Application and Benefits

The OHSAS specification is applicable to any organisation that wishes to:

- Establish an OH&S management system to eliminate or minimise risk to employees and other interested parties who may be exposed to OH&S risks associated with its activities.
- Implement, maintain and continually improve an OH&S management system.
- Assure itself of its conformance with its stated OH&S policy.
- Demonstrate such conformance to others.
- Seek certification/registration of its OH&S management system by an external organization.
- Make a self-determination and declaration of conformance with this OHSAS specification.

Steps to OHSAS 18000 Certification

The process for establishing an OHSAS 18001 management system is essentially the same as the process used of implementing an ISO 14001 system:

- Preliminary review of existing programs and systems.
- Identification of hazards and applicable laws and regulations.
- Developing new procedures.
- Training personnel.
- Implementing new programs such as internal audit and management review.
- Seeking certification of the OHS program.

The OH&S management program(s) needs to be reviewed at regular and planned intervals to address changes to the activities, products, services, or operating conditions of the organization.

Training, Awareness and Competence

The organization shall establish and maintain procedures to ensure that its employees working at each relevant function and level are aware of:

- The importance of conformance to the OH&S policy and procedures, and to the requirements of the OH&S management system;
- The OH&S consequences, actual and potential, of their work activities and the OH&S benefits of improved personal performance;
- Their roles and responsibilities in achieving conformance to the OH&S policy and procedures and to the requirements of the OH&S management system, including emergency preparedness and response requirements;
- The potential consequences of departure from specified operating procedures.

Training procedures shall take into account differing levels of: Responsibility, ability and literacy and Risk.

Employee involvement and consultation arrangements is to be documented and interested parties informed.

Employees shall be:

- Involved in the development and review of policies and procedures to manage risks;
- Consulted where there are any changes that affect workplace health and safety;
- Represented on health and safety matters; and
- Informed as to who is their employee OH&S representative(s) and specified management appointee.

Documentation

The organization needs to establish and maintain information, in a suitable medium such as paper or electronic form, that:

a) Describes the core elements of the management system and their interaction; and

b) Provides direction to related documentation;

c) Periodically reviewed, revised as necessary and approved for adequacy by authorized personnel;

d) Archival documents and data retained for legal or knowledge preservation purposes or both are suitably identified.

Operational Control

The organization shall identify those operations and activities that are associated with identified risk where control measures need to be applied. The organization shall plan these activities, including maintenance, in order to ensure that they are carried out under specified conditions by:

a) Establishing and maintaining documented procedures to cover situations where their absence could lead to deviations from the OH&S policy and the objectives;

b) Stipulating operating criteria in the procedures;

c) Establishing and maintaining procedures related to the identified OH&S risks of goods, equipment and services purchased and/or used by the organization and communicating relevant procedures and requirements to suppliers and contractors;

d) Establishing and maintaining procedures for the design of workplace, process, installations, machinery, operating procedures and work organization, including their adaptation to human capabilities, in order to eliminate or reduce OH&S risks at their source.

Emergency Preparedness and Response

The organization shall establish and maintain plans and procedures to identify the potential for, and responses to, incidents and emergency situations, and for preventing and mitigating the likely illness and injury that may be associated with them.

The organization shall review its emergency preparedness and response plans and procedures, in particular after the occurrence of incidents or emergency situations. The organization shall also periodically test such procedures where practicable.

Performance Measurement and Monitoring

The organization shall establish and maintain procedures to monitor and measure OH&S performance on a regular basis. These procedures shall provide for:

- Both qualitative and quantitative measures, appropriate to the needs of the organization;
- Monitoring of the extent to which the organization's OH&S objectives are met;
- Proactive measures of performance that monitor compliance with the OH&S management program, operational criteria and applicable legislation and regulatory requirements;
- Reactive measures of performance to monitor accidents, ill health, incidents (including near-misses) and other historical evidence of deficient OH&S performance;
- Recording of data and results of monitoring and measurement sufficient to facilitate subsequent corrective and preventive action analysis.

If monitoring equipment is required for performance measurement and monitoring, the organization shall establish and maintain procedures for the calibration and maintenance of such equipment. Records of calibration and maintenance activist and results shall be retained.

OH&S records shall be legible, identifiable and traceable to the activities involved. OH&S records shall be stored and maintained in such a way that they are readily retrievable and protected against damage, deterioration or loss. Their retention times shall be established and recorded.

Audit

The organization shall establish and maintain an audit program and procedures for periodic OH&S management system audits to be carried out, in order to:

a) Determine whether or not the OH&S management system:

i. Conforms to planned arrangements for OH&S management including the requirements of this OHSAS specification;

ii. Has been properly implemented and maintained; and

iii. Is effective in meeting the organization's policy and objectives.

b) Review the results of previous audits.

c) Provide information on the results of audits to management.

The audit program, including any schedule, shall be based on the results of risk assessments of the organization's activities, and the results of previous audits. The audit procedures shall cover the scope, frequency, methodologies and competencies, as well as the responsibilities and requirements for conducting audits and reporting results.

Wherever possible, audits shall be conducted by personnel in dependent of those having direct responsibility for the activity being examined.

The organization's top management shall, at intervals that it determines, review the OH&S management system, to ensure it continuing suitability, adequacy and effectiveness. The management review process shall ensure that the necessary information is collected to allow management to carry out this evaluation. This review shall be documented.

The management review shall address the possible need for changes to policy, objectives and other elements of the OH&S management system, in the light of OH&S management system audit results, changing circumstances and the commitment to continual improvement.

The table analyses the interrelationship among OHSAS 18001, ISO 14001 and ISO 9001:

Inter-relationship between OHSAS 18001, ISO 14001 and ISO 9001:

OHSAS 18001	ISO 14001	ISO 9001
Scope	Scope	Scope
Reference publications	Normative references	Normative references
Terms and definitions	Definitions	Definitions
OH&S management system elements	Environmental management system requirements	Quality system requirements
General requirements	General Requirements	General (1st sentence)

OH&S Policy	Environmental policy	Quality policy
Planning	Planning	Quality system
Planning for hazard identification, risk assessment and risk control	Environmental aspects	Quality system
Legal and other requirements	Legal and other requirements	--
Objectives	Objectives and targets	Quality system
OH&S management program(s)	Environmental management program(s)	Quality system
Implementation and operation	Implementation and operation	Quality system Process control
Structure and responsibility	Structure and responsibility	Management responsibility Organization
Training, awareness and competence	Training, awareness and competence	Training
Consultation and communication	Communication	--
Documentation	Environmental management system documentation	General (without 1st sentence)
Document and data control	Document control	Document and data control

Inter-relationship between OHSAS 18001, ISO 14001 and ISO 9001:

OHSAS 18001	ISO 14001	ISO 9001
Operational control	Operational control	Quality system procedures Contract review Design control Purchasing Customer supplied product Product identification and traceability Process control Handling, storage, packaging, preservation and delivery Servicing Statistical techniques
Emergency preparedness and response	Emergency preparedness and response	--
Checking and corrective action	Checking and corrective action	--
Performance measurement and monitoring	Monitoring and measurement	Inspection and testing Control of inspection, measuring and test equipment Inspection and test status

Accidents, incidents, non-conformances and corrective and preventive action	Non-conformance and corrective and preventive action	Control of non- conforming product Corrective and preventive action
Records and records management	records	Control of quality records
Audit	Environmental management system audit	Internal quality audits
Management review	Management review	Management review

BS 18004:2008

BS 18004 provides guidance for an occupational health and safety (OH&S) management system or the OH&S elements of an organization's overall management system. It enables an organization to control its OH&S risks and improve its OH&S performance.

BS 18004 applies to any organization that wishes to:

- Establish an OH&S management system to control risks to personnel and other interested parties who could be exposed to OH&S hazards associated with its activities.
- Implement, maintain and continually improve an OH&S management system.
- Demonstrate commitment to good practice, including selfregulation, and continual improvement in OH&S performance.
- Assure itself of its conformity with its stated OH&S policy and with BS OHSAS 18001 by: Making self-determination and self-declaration, or Seeking confirmation of its conformity by parties having an interest in the organization, such as customers, or Seeking confirmation of its selfdeclaration by a party external to the organization, or Seeking certification/ registration of its OH&S management system by an external organization.

All elements of BS 18004 can be incorporated into any OH&S management system and enables an organization to incorporate OH&S within its overall management system. The extent of the application will depend on such factors as the OH&S policy of the organization, the nature of its activities and the risks and complexity of its operations.

Life-cycle Assessment

Life-cycle assessment (LCA, also known as life-cycle analysis, ecobalance, and cradle-to-grave analysis) is a technique to assess environmental impacts associated with all the stages of a product's life from raw material extraction through materials processing, manufacture, distribution, use, repair and maintenance, and disposal or recycling. Designers use this process to help critique their products.

LCAs can help avoid a narrow outlook on environmental concerns by:

- Compiling an inventory of relevant energy and material inputs and environmental releases;
- Evaluating the potential impacts associated with identified inputs and releases;
- Interpreting the results to help make a more informed decision.

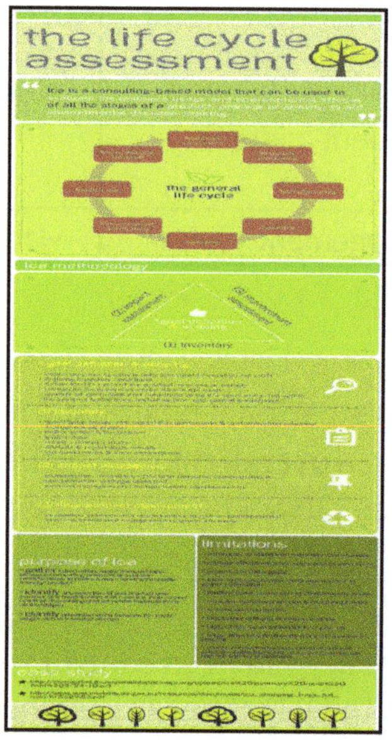

Life Cycle Assessment Overview.

Goals and Purpose

The goal of LCA is to compare the full range of environmental effects assignable to products and services by quantifying all inputs and outputs of material flows and assessing how these material flows affect the environment. This information is used to improve processes, support policy and provide a sound basis for informed decisions.

The term *life cycle* refers to the notion that a fair, holistic assessment requires the assessment of raw-material production, manufacture, distribution, use and disposal including all intervening transportation steps necessary or caused by the product's existence.

There are two main types of LCA. Attributional LCAs seek to establish (or attribute) the burdens associated with the production and use of a product, or with a specific service or process, at a point in time (typically the recent past). Consequential LCAs seek to identify the environmental consequences of a decision or a proposed change in a system under study (oriented to the future), which means that market and economic implications of a decision may have to be taken into account. Social LCA is under development as a different approach to life cycle thinking intended to assess social implications or potential impacts. Social LCA should be considered as an approach that is complementary to environmental LCA.

The procedures of life cycle assessment (LCA) are part of the ISO 14000 environmental management standards: in ISO 14040:2006 and 14044:2006. (ISO 14044 replaced earlier versions of ISO 14041 to ISO 14043.) GHG product life cycle assessments can also comply with specifications such as PAS 2050 and the GHG Protocol Life Cycle Accounting and Reporting Standard.

Four Main Phases

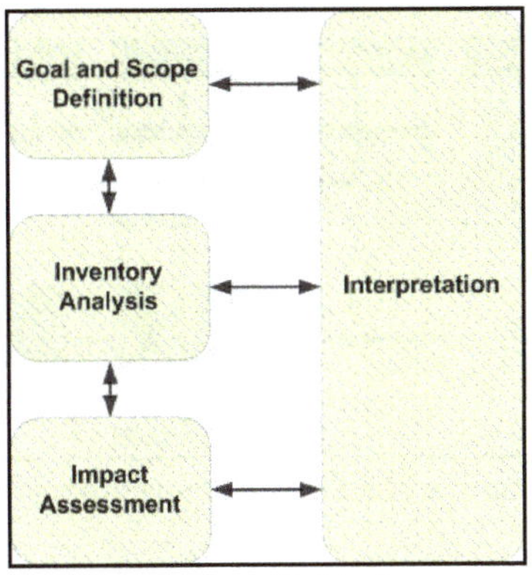

Illustration of LCA phases.

According to the ISO 14040 and 14044 standards, a Life Cycle Assessment is carried out in four distinct phases as illustrated in the figure shown to the right. The phases are often interdependent in that the results of one phase will inform how other phases are completed.

Goal and Scope

An LCA starts with an explicit statement of the goal and scope of the study, which sets out the context of the study and explains how and to whom the results are to be communicated. This is a key step and the ISO standards require that the goal and scope of an LCA be clearly defined and consistent with the intended application. The goal and scope document therefore includes technical details that guide subsequent work:

- The functional unit, which defines what precisely is being studied and quantifies the service delivered by the product system, providing a reference to which the inputs and outputs can be related. Further, the functional unit is an important basis that enables alternative goods, or services, to be compared and analyzed. So to explain this a functional system which is inputs, processes and outputs contains a functional unit, that fulfills a function, for example paint is covering a wall, making a functional unit of $1m^2$ covered for 10 years. The functional flow would be the items necessary for that function, so this would be a brush, tin of paint and the paint itself.
- The system boundaries; which are delimitations of which processes that should be included in the analysis of a product system.
- Any assumptions and limitations.
- The allocation methods used to partition the environmental load of a process when several products or functions share the same process; allocation is commonly dealt with in one of three ways: system expansion, substitution and partition. Doing this is not easy and different methods may give different results.
- The impact categories chosen for example human toxicity, smog, global warming, eutrophication.

Life Cycle Inventory

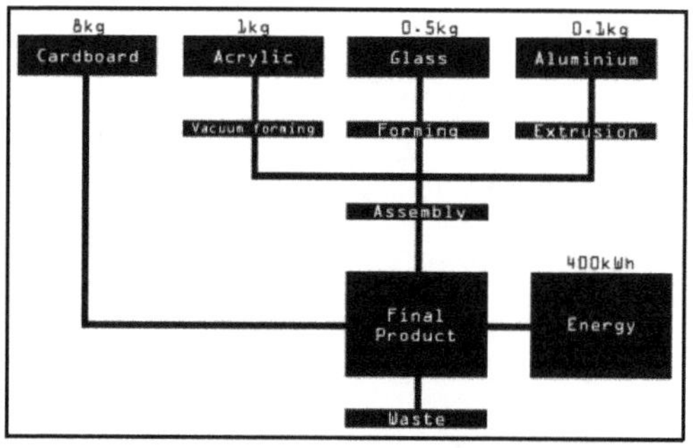

This is an example of a Life-cycle inventory (LCI) diagram.

Life Cycle Inventory (LCI) analysis involves creating an inventory of flows from and to nature for a product system. Inventory flows include inputs of water, energy, and raw materials, and releases to air, land, and water. To develop the inventory, a flow model of the technical system is constructed using data on inputs and outputs. The flow model is typically illustrated with a flow chart that includes the activities that are going to be assessed in the relevant supply chain and gives a clear picture of the technical system boundaries. The input and output data needed for the construction of the model are collected for all activities within the system boundary, including from the supply chain (referred to as inputs from the technosphere).

The data must be related to the functional unit defined in the goal and scope definition. Data can be presented in tables and some interpretations can be made already at this stage. The results of the inventory is an LCI which provides information about all inputs and outputs in the form of elementary flow to and from the environment from all the unit processes involved in the study.

Inventory flows can number in the hundreds depending on the system boundary. For product LCAs at either the generic (i.e., representative industry averages) or brand-specific level, that data is typically collected through survey questionnaires. At an industry level, care has to be taken to ensure that questionnaires are completed by a representative sample of producers, leaning toward neither the best nor the worst, and fully representing any regional differences due to energy use, material sourcing or other factors. The questionnaires cover the full range of inputs and outputs, typically aiming to account for 99% of the mass of a product, 99% of the energy used in its production and any environmentally sensitive flows, even if they fall within the 1% level of inputs.

One area where data access is likely to be difficult is flows from the technosphere. The technosphere is more simply defined as the man-made world. Considered by geologists as secondary resources, these resources are in theory 100% recyclable; however, in a practical sense, the primary goal is salvage. For an LCI, these technosphere products (supply chain products) are those that have been produced by man and unfortunately those completing a questionnaire about a process which uses a man-made product as a means to an end will be unable to specify how much of a given input they use. Typically, they will not have access to data concerning inputs and outputs for previous production processes of the product. The entity undertaking the LCA must then turn to secondary sources if it does not already have that data from its own previous studies. National databases or data sets that come with LCA-practitioner tools, or that can be readily accessed, are the usual sources for that information. Care must then be taken to ensure that the secondary data source properly reflects regional or national conditions.

LCI Methods

- Process LCA
- Economic Input Output LCA
- Hybrid Approach

Life Cycle Impact Assessment

Inventory analysis is followed by impact assessment. This phase of LCA is aimed at evaluating the significance of potential environmental impacts based on the LCI flow results. Classical life cycle impact assessment (LCIA) consists of the following mandatory elements:

- Selection of impact categories, category indicators, and characterization models;
- The classification stage, where the inventory parameters are sorted and assigned to specific impact categories; and
- Impact measurement, where the categorized LCI flows are characterized, using one of many possible LCIA methodologies, into common equivalence units that are then summed to provide an overall impact category total.

In many LCAs, characterization concludes the LCIA analysis; this is also the last compulsory stage according to ISO 14044:2006. However, in addition to the above mandatory LCIA steps, other optional LCIA elements – normalization, grouping, and weighting – may be conducted depending on the goal and scope of the LCA study. In normalization, the results of the impact categories from the study are usually compared with the total impacts in the region of interest, the U.S. for example. Grouping consists of sorting and possibly ranking the impact categories. During weighting, the different environmental impacts are weighted relative to each other so that they can then be summed to get a single number for the total environmental impact. ISO 14044:2006 generally advises against weighting, stating that "weighting, shall not be used in LCA studies intended to be used in comparative assertions intended to be disclosed to the public". This advice is often ignored, resulting in comparisons that can reflect a high degree of subjectivity as a result of weighting.

Interpretation

Life Cycle Interpretation is a systematic technique to identify, quantify, check, and evaluate information from the results of the life cycle inventory and/or the life cycle impact assessment. The results from the inventory analysis and impact assessment are summarized during the interpretation phase. The outcome of the interpretation phase is a set of conclusions and recommendations for the study. According to ISO 14040:2006, the interpretation should include:

- Identification of significant issues based on the results of the LCI and LCIA phases of an LCA;
- Evaluation of the study considering completeness, sensitivity and consistency checks; and
- Conclusions, limitations and recommendations.

A key purpose of performing life cycle interpretation is to determine the level of confidence in the final results and communicate them in a fair, complete, and accurate manner. Interpreting the results of an LCA is not as simple as "3 is better than 2, therefore Alternative A is the best choice"! Interpreting the results of an LCA starts with understanding the accuracy of the results, and ensuring they meet the goal of the study. This is accomplished by identifying the data elements that contribute significantly to each impact category, evaluating the sensitivity of these significant data elements, assessing the completeness and consistency of the study, and drawing conclusions and recommendations based on a clear understanding of how the LCA was conducted and the results were developed.

Reference Test

More specifically, the best alternative is the one that the LCA shows to have the least cradle-to-grave environmental negative impact on land, sea, and air resources.

LCA uses

Based on a survey of LCA practitioners carried out in 2006 LCA is mostly used to support business strategy (18%) and R&D (18%), as input to product or process design (15%), in education (13%) and for labeling or product declarations (11%). LCA will be continuously integrated into the built environment as tools such as the European ENSLIC Building project guidelines for buildings or developed and implemented, which provide practitioners guidance on methods to implement LCI data into the planning and design process.

Major corporations all over the world are either undertaking LCA in house or commissioning studies, while governments support the development of national databases to support LCA. Of particular note is the growing use of LCA for ISO Type III labels called Environmental Product Declarations, defined as "quantified environmental data for a product with pre-set categories of parameters based on the ISO 14040 series of standards, but not excluding additional environmental information". These third-party certified LCA-based labels provide an increasingly important basis for assessing the relative environmental merits of competing products. Third-party certification plays a major role in today's industry. Independent certification can show a company's dedication to safer and environmental friendlier products to customers and NGOs.

LCA also has major roles in environmental impact assessment, integrated waste management and pollution studies.

Data Analysis

A life cycle analysis is only as valid as its data; therefore, it is crucial that data used for the completion of a life cycle analysis are accurate and current. When comparing different life cycle analyses with one another, it is crucial that equivalent data are available for both products or processes in question. If one product has a much higher availability of data, it cannot be justly compared to another product which has less detailed data.

There are two basic types of LCA data – unit process data and environmental input-output data (EIO), where the latter is based on national economic input-output data. Unit process data are derived from direct surveys of companies or plants producing the product of interest, carried out at a unit process level defined by the system boundaries for the study.

Data validity is an ongoing concern for life cycle analyses. Due to globalization and the rapid pace of research and development, new materials and manufacturing methods are continually being introduced to the market. This makes it both very important and very difficult to use up-to-date information when performing an LCA. If an LCA's conclusions are to be valid, the data must be recent; however, the data-gathering process takes time. If a product and its related processes have not undergone significant revisions since the last LCA data was collected, data validity is not a problem. However, consumer electronics such as cell phones can be redesigned as often as every 9 to 12 months, creating a need for ongoing data collection.

The life cycle considered usually consists of a number of stages including: materials extraction, processing and manufacturing, product use, and product disposal. If the most environmentally harmful of these stages can be determined, then impact on the environment can be efficiently reduced by focusing on making changes for that particular phase. For example, the most energy-intensive life phase of an airplane or car is during use due to fuel consumption. One of the most effective ways to increase fuel efficiency is to decrease vehicle weight, and thus, car and airplane manufacturers can decrease environmental impact in a significant way by replacing heavier materials with lighter ones such as aluminium or carbon fiber-reinforced elements. The reduction during the use phase should be more than enough to balance additional raw material or manufacturing cost.

Data sources are typically large databases, it is not appropriate to compare two options if different data sources have been used to source the data. Data sources include:

- soca
- EuGeos' 15804-IA
- NEEDS
- ecoinvent
- PSILCA
- ESU World Food
- GaBi
- ELCD
- LC-Inventories.ch
- Social Hotspots
- ProBas
- bioenergiedat
- Agribalyse
- USDA
- Ökobaudat

Calculations for impact can then be done by hand, but it is more usual to streamline the process by using software. This can range from a simple spreadsheet, where the user enters the data manually to a fully automated program, where the user is not aware of the source data.

Variants

Cradle-to-grave

Cradle-to-grave is the full Life Cycle Assessment from resource extraction ('cradle') to use phase and disposal phase ('grave'). For example, trees produce paper, which can be recycled into low-energy production cellulose (fiberised paper) insulation, then used as an energy-saving device in the ceiling of a home for 40 years, saving 2,000 times the fossil-fuel energy used in its production. After 40 years the cellulose fibers are replaced and the old fibers are disposed of, possibly incinerated. All inputs and outputs are considered for all the phases of the life cycle.

Cradle-to-gate

Cradle-to-gate is an assessment of a *partial* product life cycle from resource extraction (*cradle*) to the factory gate (i.e., before it is transported to the consumer). The use phase and disposal phase of the product are omitted in this case. Cradle-to-gate assessments are sometimes the basis for environmental product declarations (EPD) termed business-to-business EDPs. One of the significant uses of the cradle-to-gate approach compiles the life cycle inventory (LCI) using cradle-to-gate. This allows the LCA to collect all of the impacts leading up to resources being purchased by the facility. They can then add the steps involved in their transport to plant and manufacture process to more easily produce their own cradle-to-gate values for their products.

Cradle-to-cradle or Closed Loop Production

Cradle-to-cradle is a specific kind of cradle-to-grave assessment, where the end-of-life disposal step for the product is a recycling process. It is a method used to minimize the environmental impact of products by employing sustainable production, operation, and disposal practices and aims to incorporate social responsibility into product development. From the recycling process originate new, identical products (e.g., asphalt pavement from discarded asphalt pavement, glass bottles from collected glass bottles), or different products (e.g., glass wool insulation from collected glass bottles).

Allocation of burden for products in open loop production systems presents considerable challenges for LCA. Various methods, such as the avoided burden approach have been proposed to deal with the issues involved.

Gate-to-gate

Gate-to-gate is a partial LCA looking at only one value-added process in the entire production chain. Gate-to-gate modules may also later be linked in their appropriate production chain to form a complete cradle-to-gate evaluation.

Well-to-wheel

Well-to-wheel is the specific LCA used for transport fuels and vehicles. The analysis is often broken down into stages entitled "well-to-station", or "well-to-tank", and "station-to-wheel" or "tank-to-wheel", or "plug-to-wheel". The first stage, which incorporates the feedstock or fuel production and processing and fuel delivery or energy transmission, and is called the "upstream" stage, while the stage that deals with vehicle operation itself is sometimes called the "downstream" stage. The well-to-wheel analysis is commonly used to assess total energy consumption, or the energy conversion efficiency and emissions impact of marine vessels, aircraft and motor vehicles, including their carbon footprint, and the fuels used in each of these transport modes.

The well-to-wheel variant has a significant input on a model developed by the Argonne National Laboratory. The Greenhouse gases, Regulated Emissions, and Energy use in Transportation (GREET) model was developed to evaluate the impacts of new fuels and vehicle technologies. The model evaluates the impacts of fuel use using a well-to-wheel evaluation while a traditional cradle-to-grave approach is used to determine the impacts from the vehicle itself. The model reports energy use, greenhouse gas emissions, and six additional pollutants: volatile organic compounds (VOCs), carbon monoxide (CO), nitrogen oxide (NOx), particulate matter with size smaller than 10 micrometre (PM10), particulate matter with size smaller than 2.5 micrometre (PM2.5), and sulfur oxides (SOx).

Economic Input–output Life Cycle Assessment

Economic input–output LCA (EIOLCA) involves use of aggregate sector-level data on how much environmental impact can be attributed to each sector of the economy and how much each sector purchases from other sectors. Such analysis can account for long chains (for example, building an automobile requires energy, but producing energy requires vehicles, and building those vehicles requires energy, etc.), which somewhat alleviates the scoping problem of process LCA; however, EIOLCA relies on sector-level averages that may or may not be representative of the specific subset of the sector relevant to a particular product and therefore is not suitable for evaluating the environmental impacts of products. Additionally the translation of economic quantities into environmental impacts is not validated.

Ecologically Based LCA

While a conventional LCA uses many of the same approaches and strategies as an Eco-LCA, the latter considers a much broader range of ecological impacts. It was designed to provide a guide to wise management of human activities by understanding the direct and indirect impacts on ecological resources and surrounding ecosystems. Developed by Ohio State University Center for resilience, Eco-LCA is a methodology that quantitatively takes into account regulating and supporting services during the life cycle of economic goods and products. In this approach services are categorized in four main groups: supporting, regulating, provisioning and cultural services.

Exergy Based LCA

Exergy of a system is the maximum useful work possible during a process that brings the system into equilibrium with a heat reservoir. Wall clearly states the relation between exergy analysis and resource accounting. This intuition confirmed by DeWulf and Sciubba lead to Exergo-economic accounting and to methods specifically dedicated to LCA such as Exergetic material input per unit of service (EMIPS). The concept of material input per unit of service (MIPS) is quantified in terms of the second law of thermodynamics, allowing the calculation of both resource input and service output in exergy terms. This exergetic material input per unit of service (EMIPS) has been elaborated for transport technology. The service not only takes into account the total mass to be transported and the total distance, but also the mass per single transport and the delivery time.

The applicability of the EMIPS methodology relates specifically to transport system. This model has been further improved by Trancossi who has introduced the friction term, which has not been considered by original EMIPS model, and the key distinction between exergy disruption by payload and by vehicle, focusing on the losses due to vehicle and more effective evaluation of the processes and produced an effective assessment of today transport vehicles. This model is referenced by Indian "Road less traveled" model, which has been developed for minimizing the impact of transports in urban environment.

Life Cycle Energy Analysis

Life cycle energy analysis (LCEA) is an approach in which all energy inputs to a product are accounted for, not only direct energy inputs during manufacture, but also all energy inputs needed to produce components, materials and services needed for the manufacturing process. An earlier term for the approach was *energy analysis*.

With LCEA, the *total life cycle energy input* is established.

Energy Production

It is recognized that much energy is lost in the production of energy commodities themselves, such as nuclear energy, photovoltaic electricity or high-quality petroleum products. *Net energy content* is the energy content of the product minus energy input used during extraction and conversion, directly or indirectly. A controversial early result of LCEA claimed that manufacturing solar cells requires more energy than can be recovered in using the solar cell. The result was refuted. Another new concept that flows from life cycle assessments is Energy Cannibalism. Energy Cannibalism refers to an effect where rapid growth of an entire energy-intensive industry creates a need for energy that uses (or cannibalizes) the energy of existing power plants. Thus during rapid growth the industry as a whole produces no energy because new energy is used to fuel the embodied energy of future power plants. Work has been undertaken in the UK to determine the life cycle energy (alongside full LCA) impacts of a number of renewable technologies.

Energy Recovery

If materials are incinerated during the disposal process, the energy released during burning can be harnessed and used for electricity production. This provides a low-impact energy source, especially when compared with coal and natural gas While incineration produces more greenhouse gas emissions than landfilling, the waste plants are well-fitted with filters to minimize this negative impact. A recent study comparing energy consumption and greenhouse gas emissions from landfilling (without energy recovery) against incineration (with energy recovery) found incineration to be superior in all cases except for when landfill gas is recovered for electricity production.

Criticism

A criticism of LCEA is that it attempts to eliminate monetary cost analysis, that is replace the currency by which economic decisions are made with an energy currency. It has also been argued that energy efficiency is only one consideration in deciding which alternative process to employ, and that it should not be elevated to the only criterion for determining environmental acceptability; for example, simple energy analysis does not take into account the renewability of energy flows or the toxicity of waste products; however the life cycle assessment does help companies become more familiar with environmental properties and improve their environmental system. Incorporating Dynamic LCAs of renewable energy technologies (using sensitivity analyses to project future improvements in renewable systems and their share of the power grid) may help mitigate this criticism.

In recent years, the literature on life cycle assessment of energy technology has begun to reflect the interactions between the current electrical grid and future energy technology. Some papers have focused on energy life cycle, while others have focused on carbon dioxide (CO_2) and other greenhouse gases. The essential critique given by these sources is that when considering energy technology, the growing nature of the power grid must be taken into consideration. If this is not done, a given class of energy technology may emit more CO_2 over its lifetime than it mitigates.

A problem the energy analysis method cannot resolve is that different energy forms (heat, electricity, chemical energy etc.) have different quality and value even in natural sciences, as a consequence of the two main laws of thermodynamics. A thermodynamic measure of the quality of energy is exergy. According to the first law of thermodynamics, all energy inputs should be accounted with equal weight, whereas by the second law diverse energy forms should be accounted by different values.

The conflict is resolved in one of these ways:

- Value difference between energy inputs is ignored,
- A value ratio is arbitrarily assigned (e.g., a joule of electricity is 2.6 times more valuable than a joule of heat or fuel input),
- The analysis is supplemented by economic (monetary) cost analysis,
- Exergy instead of energy can be the metric used for the life cycle analysis.

Critiques

Life cycle assessment is a powerful tool for analyzing commensurable aspects of quantifiable systems. Not every factor, however, can be reduced to a number and inserted into a model. Rigid system boundaries make accounting for changes in the system difficult. This is sometimes referred to as the boundary critique to systems thinking. The accuracy and availability of data can also contribute to inaccuracy. For instance, data from generic processes may be based on averages, unrepresentative sampling, or outdated results. Additionally, social implications of products are generally lacking in LCAs. Comparative life-cycle analysis is often used to determine a better process or product to use. However, because of aspects like differing system boundaries, different statistical information, different product uses, etc., these studies can easily be swayed in favor of one product or process over another in one study and the opposite in another study based on varying parameters and different available data. There are guidelines to help reduce such conflicts in results but the method still provides a lot of room for the researcher to decide what is important, how the product is typically manufactured, and how it is typically used.

An in-depth review of 13 LCA studies of wood and paper products found a lack of consistency in the methods and assumptions used to track carbon during the product lifecycle. A wide variety of methods and assumptions were used, leading to different and potentially contrary conclusions – particularly with regard to carbon sequestration and methane generation in landfills and with carbon accounting during forest growth and product use.

Streamline LCA

This process includes three steps. First, a proper method should be selected to combine adequate accuracy with acceptable cost burden in order to guide decision making. Actually, in LCA process, besides streamline LCA, Eco-screening and complete LCA are usually considered as well. However, the former one only could provide limited details and the latter one with more detailed information is more expensive. Second, single measure of stress should be selected. Typical LCA output includes resource consumption, energy consumption, water consumption, emission of CO_2, toxic residues and so on. One of these outputs is used as the main factor to measure in streamline LCA. Energy consumption and CO_2 emission are often regarded as "practical indicators". Last, stress selected in step 2 is used as standard to assess phase of life separately and identify the most damaging phase. For instance, for a family car, energy consumption could be used as the single stress factor to assess each phase of life. The result shows that the most energy intensive phase for a family car is the usage stage.

Life Cycle Assessment of Engineered Material in Service plays a significant role in saving energy, conserving resources and saving billions by preventing premature failure of critical engineered component in a machine or equipment. LCA data of surface engineered materials are used to improve life cycle of the engineered component. Life cycle improvement of industrial machineries and equipments including, manufacturing, power generation, transportations, etc. leads to improvement in energy efficiency, sustainability and negating global temperature rise. Estimated reduction in anthropogenic carbon emission is minimum 10% of the global emission.

Life Cycle Assessment and its Purpose

Life cycle assessment (LCA) is a tool to evaluate the environmental effects of a product or process throughout its entire life cycle. An LCA entails examining the product from the extraction of raw materials for the manufacturing process, through the production and use of the item, to its final disposal, and thus encompassing the entire product system.

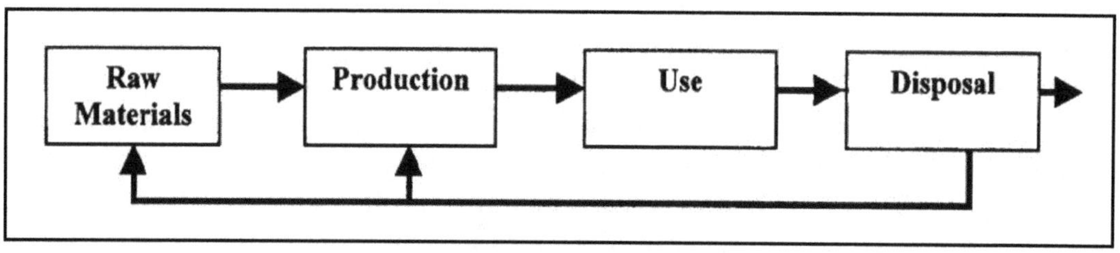

Product Life Cycle.

The assessment process includes identifying and quantifying energy and materials used and wastes released to the environment, assessing their environmental impact and evaluating opportunities for improvement.

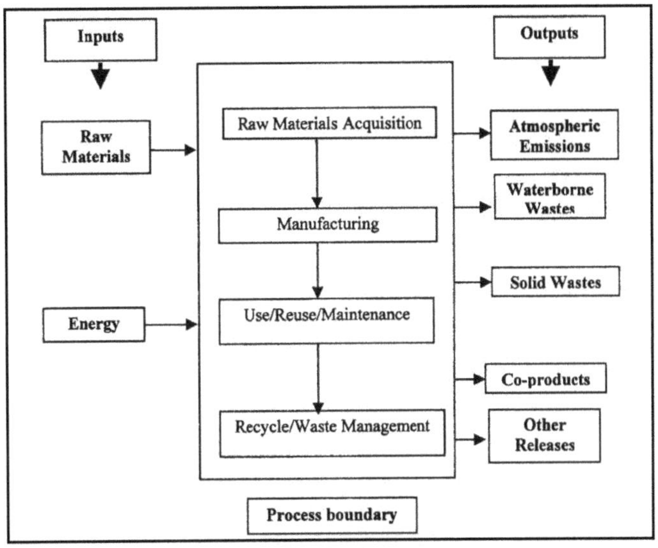

LCA Process.

The unique feature of this type of assessment is its focus on the entire life cycle, rather than a single manufacturing step or environmental emission. The theory behind this approach is that operations occurring within a facility can also cause impacts outside the facility's gates that need to be considered when evaluating project alternatives.

Purpose of LCA

Government and customers who purchase products from different sectored companies are keen on environmental properties of all the products. EMAS and ISO 14000 are demanding continual improvement in the process of production and in the environmental management system. At this stage there is a need for a tool like LCA that helps organisations meet the demand to improvise process/product.

Evolution oF Life Cycle Assessment

The principles underlying an LCA were developed in the late 1960s. In the 1970's, the US Environmental protection agency refined the methodology for evaluation of environmental impacts of products and were popularly known as resource and environmental profile analysis (REPA). Initially, it was used mainly on the consumption of energy and other resources. Knowledge of environmentally damaging releases and actions and the estimation of their effects, was too rudimentary at that time to allow a quantitative treatment of the environmental impacts of the product life cycle.

Assessments of product life cycle experienced a renaissance through studies of the environmental loadings and potential impacts of beverage containers (e.g., beer cans, milk containers) performed in various European countries in the early 1980s. These studies involved further elaborations of the principles underlying the assessment of product life cycle and entailed a series of life cycle assessments of materials used in packaging containers (i.e., polyethylene, cardboard, aluminium, etc.). A common feature of the items analysed was their homogenous character and their widespread use in many different contexts.

The late 1980s and early 1990s have seen international attempts to standardise the principles underlying life cycle assessments and to develop codes of good conduct in this field. The list of products that have been subjected to LCA has grown quickly and now includes more complex products such as paints, insulation materials, window frames, refrigerators, hotplates, television sets, etc., as well as the entire service systems or technologies such as electricity production.

As a part of ISO 14000 series of standards ISO in 2000 has come out with the following standards:

- ISO 14040: Environmental Management – LCA – Principles and Framework.
- ISO 14041: Environmental Management – LCA – Inventory Analysis.
- ISO 14042: Environmental Management – LCA – Impact Assessment.
- ISO 14043: Environmental Management – LCA – Interpretation.

Since the last decade or so, LCA is gaining importance as an environmental management tool. It has now emerged as a decision support tool in such areas as business, regulation and policy and to structure technology development in a coherent way. Many potential applications of LCA are envisaged including product improvement and design, environmental management, eco-labelling, green accounting, environmental auditing and reporting, resource management, definition of best available technology (BAT), product policy, strategic industrial planning, strategic environmental policy development, etc. As a general concept, the life cycle approach aims to support the overall goal of sustainability.

Stages in Product LCA

LCA is split into five stages that include:

1. Planning : Includes Statement of objectives, Definition of the product and its alternatives, Choice of system boundaries, Choice of environmental parameters, Choice of aggregation and evaluation method and Strategy for data collection.

2. Screening: Includes preliminary execution of LCA and adjustment of plan.

3. Data collection and treatment: Includes measurements, interviews, literature search, theoretical calculations, database search, qualified guessing and also computation of the inventory table.

4. Evaluation: Includes classification of inventory table into impact categories, aggregation within category, normalisation and weighting of different categories.

5. Improvement assessment: Includes sensitivity analysis and improvement priority and feasibility assessment.

It is generally recognized that the first stage is extremely important. The result of the LCA is heavily dependent on the decisions taken in this phase.

The screening LCA is a useful step to check the goal-definition phase. After screening it is much easier to plan the rest of the project.

The following figure gives a diagrammatic representation of the stages in the life cycle of a product:

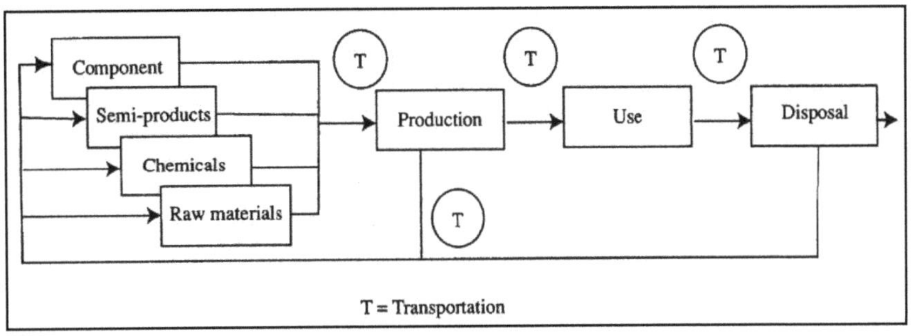

Product Life Cycle Stages.

As depicted in these figures, each stage of the life cycle receives materials and energy as inputs and produces:

- Outputs of material or products to subsequent life cycle stages;
- Emissions to the environment.

There are a number of issues associated with the life cycle stages, and we will touch upon some of these, next.

Extraction of Raw Materials

This stage in the life cycle includes the extraction of all materials involved in the entire life cycle of the product. Typical examples of activities included in this stage are forest logging and crop harvesting, fishing and mining of ores and minerals. The inventory for the extraction of raw materials should include raw materials for the production of the machinery (i.e., capital equipment) involved in manufacturing the product and other stages of the product life cycle. Raw materials used in the production of electricity and energy used in the different life stages of the product should also be considered. Collecting data for the raw materials extraction stage may prove to be a complex task. It may also lead into iterative processes such as assessing the inputs and outputs related to extraction of the raw materials that is used in the production of end products. Often, the most serious environmental problems of the product life cycle associated with this first stage. It is a common error to leave out parts of the raw materials stage from the LCA. Essentially, the decision of what to include or exclude in the LCA should be based on a sensitivity analysis.

Manufacture of a Product

The manufacturing stage encompasses all the processes involved in the conversion of raw materials into the products considered in the LCA. Apart from the manufacturing processes at the plant where the product is made, this stage takes into account production of ancillary materials, chemicals and specific or general components at other plants, no matter where they are.

Transportation

Transportation is really not a single life stage in itself. Rather, it is an integral part of all stages of the life cycle. Transportation could be characterised as conveyance of materials or energy between different operations at various locations. Included in this stage, apart from the transport process itself, is the production of packaging materials for the transportation of the product. The transport stage would possibly also include an appropriate share of the environmental loadings and consumptions associated with the construction and maintenance of the transport system, whether this is road, rail, water or air transportation.

Use of Product

The use-stage of the product occurs when it is put in service and operated over its useful life. This begins after the distribution of the product and ends when the product is used up or discarded to the waste management system. Included in the use-stage are releases and resource consumptions created by the use or maintenance of the product.

Waste Management

Wastes are generated in each phase of the life cycle, and they need to be properly managed to protect the environment. The management of wastes may involve alternative processes such as the following:

(i) Reuse: This means the use of the product or parts thereof in new units of the same product or in different products.

(ii) Recycling: This means the use of materials in the product for manufacture of the same or other products.

(iii) Incineration: This refers to the combustion of the product, generating heat that may be used for electricity production or heating.

(iv) Composting: This refers to the microbial degradation of biological materials yielding compost for improvement of agricultural soils.

(v) Waste water treatment: This refers to the organic matter degradation and nutrients removal from sewage water, creating sludge that is deposited on agricultural land.

(vi) Land filling: This means the deposition of the product in landfills.

Each form of waste treatment mentioned above may be considered a processing of waste associated with a certain consumption of resources. This results in various releases into the environment, and the possible generation of energy or materials that will be an input to the manufacturing process of this product or of other products. Before you read any further, it is a good idea to look at the Course Municipal Solid Waste Management in which we discussed the different waste management processes in detail.

As implied, the LCA of products is an important environmental management tool. However, as with every tool, difficulties do arise with LCA too.

A Code of Good Conduct for LCA

As mentioned earlier, LCA emerged during the last decade as a tool to provide an objective assessment of the total environmental impact associated with a product through its entire life cycle. In several countries, LCA is considered the primary tool by which environmental impacts of products should be regulated by government authorities. However, problems do exist. Let us look at some of the major LCA limitations below, as revealed in various studies:

(i) Data quality: In a manufacturer-sponsored study to compare a product with its alternatives, the consultants performing the LCA were able to get a very detailed and current data from the manufacturer for the processes involved in the production of that product. However, they had to depend on secondary data from the literature or earlier studies with regard to the production processes concerning the alternative products. Obviously, comparative studies on the basis of the secondary data tend to lack credibility.

(ii) Life cycle boundaries: A Dutch study excluded the production of several raw materials, including crude oil, for the polycarbonate production. The German study did not include emissions from the energy production associated with the life cycle of the milk containers. Most studies did not consider the working environment.

(iii) Country-specific technology types: An example is the LCA conducted on the production of electricity used in the product life cycle. In a Swedish study, the electricity production was based on nuclear power and hydroelectricity, while a Swiss study, based on a US energy scenario since 1972, used coal as the energy source. This should explain the difference in the emissions found in the two studies. In addition, as the waste processing systems contribute to the nature and level of emissions, it is important to examine these systems as well. For example, if the waste is incinerated and the combustion energy is used for electricity production, the Dutch study may tend to favour carton containers over recyclable polycarbonate bottles.

(iv) Evaluation stage priorities: In the cases where alternatives do not have particular advantage over one another, the priorities in the evaluation stage become decisive for the outcome. A recommendation of the Danish study (referred to at (ii) above) was based on an evaluation stage choice between reduced consumption of raw materials and water, but on the releases of dichloromethane into the working environment (for polycarbonate bottles), and higher energy consumption and loading of bio-accumulating and possibly carcinogenic chlorinated compounds in the waste from production (for the milk cartons).

Considering the versatility and many diverse possible applications of the LCA tool, it may be difficult to obtain reproducible and consistent results through standardisation, without losing the necessary flexibility for adaptation to particular cases being studied. Given this scenario, it is important that we include the following factors in a code of good conduct for LCA:

(v) Definition: This involves finding answers to such questions as: Is the purpose of the study explicitly defined? Is it meant for internal company use or for public use? If the study is intended for public use, has it been peer reviewed? Is it clearly stated for whom the study is performed and by whom it is sponsored? Is the definition of the functional unit appropriate?, etc.

(vi) Delimitation of system under study: This involves finding answers to such questions as: Is there an explicit and clear delimitation of the system under study? Is the life cycle described in detail, stage by stage? Are the life cycle description and process tree plausible? Do they describe the real world system in a realistic way? Does the study include the extraction of raw materials?

Does the study include the production of electricity? What production scenario is used? Is it appropriate? Does the study include the manufacture of real capital for all life cycle stages? If the study has precluded capital, is this omission substantiated? Do you find the omission reasonable? Is the disposal stage covered by the study?, etc.

(vii) Inventory: This involves finding answers to such questions as: Does the inventory cover all processes of the process tree? Is there a reference to the source of every piece of data in the inventory? Is the data quality appropriate, i.e., primary and recent data for all the important processes of the life cycle? Do the data describe a relevant technological level of the processes? Is the use of data of lower quality or omission of processes from the inventory based on sensitivity analyses? Is the quality of the data used for the compared alternatives similar? Are data on impacts to the working environment present in the inventory? Is it acceptable that they be left out in the considered study?, etc.

(viii) Impact assessment: This involves finding answers to such questions as: Has any impact assessment been performed? Does it consider all the important environmental effect types? Does it consider resource and working environment issues?, etc.

(ix) Evaluation: This involves finding answers to such questions as: Are the subjective steps of the evaluation separated from the objective ones? If not, is the evaluation transparent? Are the priorities clear? Are the conclusions of the assessment clear? Are the assumptions underlying the weighting explained, and do you agree with them?, etc.

Procedure for LCA

Four steps are involved in carrying out an LCA, and these are:

(i) Definition of scope, goals, and delimitation of the life cycle.

(ii) Preparation of an inventory.

(iii) Assessment of impact of environmental loadings in terms of environmental profiles.

(iv) Evaluation of environmental profiles according to the defined goals.

Defining the Goal and Scope

The goal of an LCA study must unambiguously state the intended application, including the reasons for carrying out the study and the intended audience. The statement of goal must also indicate the intended use of the results and users of the results. The practitioner, who has to reach the goal, needs to understand the detailed purpose of the study in order to make proper decisions throughout the study. Examples of goals of life cycle assessments are:

- To compare two or more different products fulfilling the same function with the purpose of using the information in marketing of the products or regulating the use of the products.
- To identify improvement possibilities in further development of existing products or in innovation and design of new products.
- To identify areas, steps, etc., in the life cycle of a product to meet the eco-labelling criteria.

Transparency is essential for all kinds of LCA studies. The target group of the LCA study is also important to have in mind the choice of reporting method. The goal can be redefined as a result of the findings throughout the study.

While the goal definition determines the level of sophistication of the study and the requirements to reporting, the definition of the scope of an LCA sets the borders of the assessment, i.e., what may be included in the system and what detailed assessment methods are to be used.

In defining the scope of an LCA study, the following elements must be considered and clearly described:

- The functions of the system, or in the case of comparative studies, systems.
- The functional unit.
- The system to be studied.
- The system boundaries.
- The allocation procedures.
- The types of impact and the methodology of impact assessment and subsequent interpretation to be used.
- Data requirements.
- Assumptions.
- Limitations.
- The initial data quality requirements.
- The type of critical review, if any.
- The type and format of the report required for the study.

The scope should be sufficiently well defined to ensure that the breadth, the depth and the detail of the study are compatible and sufficient to address the stated goal. LCA is an iterative technique.

Therefore, the scope of the study may need to be modified while the study is being conducted as additional information is collected.

Let us now describe some of the elements mentioned above to define the scope of an LCA.

Functional Unit

Definition of the functional unit or performance characteristics is the foundation of an LCA because the functional unit sets the scale for comparison of two or more products including improvement to one product (system). All data collected in the inventory phase will be related to the functional unit. When comparing different products fulfilling the same function, definition of the functional unit is of particular importance. The three aspects that have to be taken into account when defining a functional unit are the efficiency of the product, the durability of the product and the performance quality standard (Lindfors et al., 1995c).

When performing an assessment of more complicated systems, e.g., multi-functional systems, special attention has to be paid to by-products. Waste treatment systems are an example of processes with different outputs (e.g., energy, a fertiliser). When comparing different systems, inclusion of the produced amount of energy and fertiliser is an example of handling of different byproducts in the definition of the functional unit.

System Boundaries

The system boundaries define the processes/operation (e.g., manufacturing, transport and waste management processes) and the inputs and outputs to be taken into account in the LCA. The input can be the overall input to a production as well as input to a single process, and the same is true for the output. The definition of system boundaries is a quite subjective operation and includes geographical boundaries, life cycle boundaries (i.e., limitations in the life cycle) and boundaries between the technosphere and biosphere. Due to the subjectivity of definition of system boundaries, transparency of the defining process and the assumptions are extremely important. Note that wastewater treatment is an example of a process that often is omitted when defining the system boundaries.

Data Quality

The quality of the data used in the life cycle inventory is naturally reflected in the quality of the final LCA. The data quality can be described and assessed in different ways. It is important that the data quality is described and assessed in systematic ways that allows others to understand and control the actual data quality. Further descriptions, which define the nature of the data collected from specific sites and that from published sources and indicate whether the data should be measured, calculated or estimated, should also be considered for maintaining data quality.

In all studies, however, the following additional data quality indicators are to be taken into consideration at a certain level of detail, depending on goal and scope definition:

- Precision: That is, the measure of variability of the data values for each category expressed (e.g., variance).
- Completeness: That is, the percentage of locations reporting primary data from the potential number in existence for each data category in a unit process.
- Representativeness: This means the qualitative assessment of the degree to which the data set reflects the true population of interest (i.e., geographic, time period and technology coverage).
- Reproducibility: This means the qualitative assessment of the extent to which formation about the methodology and data values allows an independent practitioner to reproduce the results reported in the study.
- Consistency: This means the qualitative assessment of how uniformly the study methodology is applied to the various components of the analysis.

Critical Review Process

The purpose of the critical review process is to ensure the quality of the life cycle assessment. The review can be either internal or external. This may also involve the interested parties as defined within the goal and scoping definition. The critical review process ensures that:

- The methods used to carry out the LCA are consistent with the international standard and are scientifically and technically valid;
- The data used are appropriate and reasonable in relation to the goal of the study;
- The interpretations reflect the limitations identified and the goal of the study;
- The study report is transparent and consistent.

Analysing the Inventory

Inventory analysis is the second phase in an LCA, consisting of issues such as data collection, refining system boundaries, calculation, verification of data, relating data to the specific system and allocation.

We will describe each of these issues, next.

Data Collection

The inventory analysis includes collection and treatment of data to be used in preparation of a material consumption, waste and emission profile for the phases in each life cycle. The data can be both site-specific (e.g., from specific companies, specific areas and specific countries) and general sources (e.g., trade organisations, public surveys, etc.) The data have to be collected from all single processes in the life cycle. The quantitative data are important in comparisons of processes or materials, but often the quantitative data are missing or the quality is poor (too old or not technically representative, etc.). However, a more descriptive qualitative data can be used for environmental aspects or single steps in the life cycle that cannot be quantified. This can be used when the goal and scope definitions allow a non-quantitative description of the conditions.

Data collection is often the most work intensive part of a life cycle assessment, especially if site-specific data are required for all the single processes in the life cycle. In many cases, average data from the literature (often previous investigations of the same or similar products or materials) or data from trade organisations are used. The average data can be used in the conception or simplified LCA to get a first impression of the potential inputs and outputs from producing specific materials. When doing a detailed LCA, site-specific data is preferred. Note that since average data are often some years old, they may not represent the latest in technological development.

Refining System Boundaries

The system boundaries are defined as part of the scope definition procedure. After the initial data collection, the system boundaries can be refined, i.e., as a result of decisions of exclusion of life stages or sub-systems, exclusion of material flows or inclusion of new unit processes shown to be significant, according to the sensitivity analysis. The results of this refining process and the sensitivity analysis are to be documented. This analysis serves to limit the subsequent data handling to those inputs and output data, which are determined to be significant to the goal of the LCA study.

Calculation Procedures

No formal demand exists for calculation in life cycle assessment except the described demands for allocation procedures. Due to the amount of data, it is recommended, as a minimum, to develop a spreadsheet for the specific purpose. The appropriate programme can be chosen, depending on the kind and amount of data to be handled.

Validation of Data

The validation of data has to be conducted during the data collection process in order to improve the overall data quality. Systematic data validation may point out the areas where data quality must be improved or data must be found in the similar processes. During the process of data collection, a permanent and iterative check on data validity should be conducted. Validation may involve establishing mass balances, energy balances and/or comparative analyses of emission factors. Obvious anomalies in the data appearing from such validations will result in alternative data values complying with the data quality requirements.

For each data category and for each reporting location where missing data are identified, the treatment of the missing data should result in an acceptable reported data value, a zero data value if justified and a calculated value, based on the reported value from unit processes employing similar technology.

Relating Data

The fundamental input and output data are often delivered from industry in arbitrary units, e.g., energy consumption as MJ/machine/week emissions to the sewage system as mg metals/liter wastewater. The specific machine or wastewater stream is rarely connected to the production of the considered product alone but often to a number of similar products or perhaps to the whole production activity.

For each unit process, an appropriate reference flow has to be determined (e.g., one kilogram of material, one mega joule for energy, etc.). The quantitative input and output data of the unit process shall be calculated in relation to this reference flow. Based on the refined flow chart and systems boundary, unit processes are interconnected to allow calculations of the complete system. The calculation should result in all system input and output data being referenced to the functional unit. Care should be taken when aggregating the inputs and the outputs in the product system. The level of aggregation should be sufficient to satisfy the goal of the study.

Allocation and Recycling

When performing a life cycle assessment of a complex system, it may not be possible to handle all the impacts and outputs inside the system boundaries. This problem can be solved through either of the following ways:

- Expanding the system boundaries to include all the inputs and outputs;
- Allocating the relevant environmental impacts to the studied system.

When avoiding allocation by expanding the system boundaries, there is a risk of making the system too complex. The data collection, impact assessment and interpretation can then become too expensive and unrealistic in terms of time and money. Allocation may be a better alternative, if an appropriate method can be found for solving the actual problem. Allocation can be necessary when dealing with:

- Multi-output black box processes, i.e., when more than one product is produced and some of those product flows are crossing the system boundaries;
- Multi-input processes, such as waste treatment, where a strict quantitative causality between inputs and emissions seldom exists;
- Open-loop recycling, where a waste material leaving the system boundaries is used as a raw material by another system, outside the boundaries of the studied system.

On the basis of the principles presented above, the following descending order of allocation procedures is generally recommended:

- Wherever possible, allocation should be avoided or minimised. This may be achieved by subdividing the unit process into two or more sub-processes, some of which can be excluded from the system under study. Transport and materials handling are examples of processes, which can sometimes be partitioned in this way.
- Where allocation cannot be avoided, the system inputs and outputs should be partitioned among their different products or functions in such a way as to reflect the underlying physical relationship among them. That is to say, they must reflect the way in which the inputs and outputs are changed by quantitative changes in the products or functions delivered by the system.
- Where physical relationship cannot be established or used as the basis for allocation, the inputs should be allocated between the products and functions in such a way as to reflect economic relationships between them. For example, burdens might be allocated between co-products in proportion to the economic value of the products.

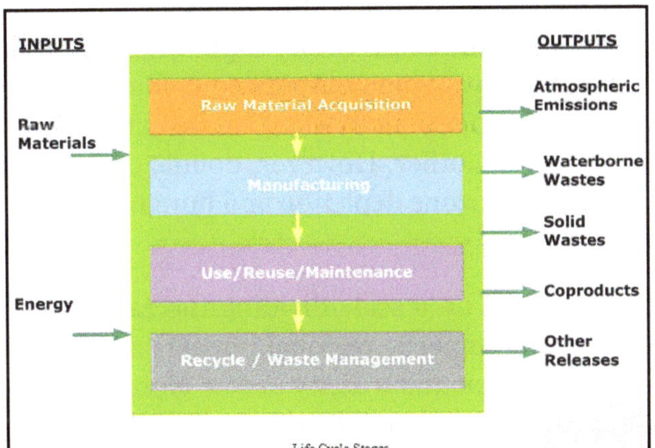

Life Cycle Stages

Assessing Environmental Impact

Impact assessment involves category definition, classification, characterisation and valuation/weighting. Let us now discuss each of these elements.

Category Definition

The life cycle assessment involves, as a first element, the definition of the impact categories to be considered (ISO, 1997c). This is a follow-up of the decisions made in the goal and scoping phase. Based on the type of information collected in the inventory phase, however, the boundaries defined in the goal and scoping may be redefined.

The impact categories are selected in order to describe the impacts caused by the products or product systems considered. The issues that need to be considered when selecting impact categories include the following:

- Completeness: This means that all environmental problems of relevance should be covered by the list.
- Practicality: This means that the list should not contain too many categories.
- Independence: This means that double counting should be avoided by choosing mutually independent impact categories.
- Relation to the characterisation step: This means that the chosen impact categories should be related to available characterisation methods (Lindfors et. al., 1995).

The impact categories considered are abiotic resources, biotic resources, land use, global warming, stratospheric ozone depletion, ecotoxicological impacts, human toxicological impacts, photochemical oxidant formation, acidification, eutrophication and work environment.

Classification

The life cycle impact assessment includes, as a second element, classification of the inventory input and output data (ISO, 1997c).

Classification is a qualitative step based on a scientific analysis of the relevant environmental processes. The classification has to assign the inventory input and output data to potential environmental impacts, i.e., impact categories. Some outputs contribute to more than one impact category, and therefore, they have to be mentioned twice. The resulting double counting is acceptable if the effects are independent of each other. However, double counting of different effects in the same effect chain (e.g., stratospheric ozone depletion and human toxicological effects, such as skin cancer) is not allowed.

The impact categories can be placed on a scale dividing the categories into four different space groups: global impacts, continental impacts, regional impacts and local impacts. The grouping is not mutually exclusive (e.g., environmental toxicity can be global, continental, regional as well as local.) The impact categories are often related directly to exposure (e.g., global exposure leads to continental impacts). Some of the impact categories are strongly correlated with continental, regional or local conditions. Certain lakes in Scandinavia can be mentioned as examples of localities that are more predisposed to acidification than lakes in other parts of Europe. The time aspect is also important when considering certain impact categories (e.g., global warming and stratospheric ozone depletion with time horizons of 20 to 500 years).

Now, let us consider an example of classification. In the manufacture of refrigerators, many of the activities relating to the production process involve the combustion of fossil fuels (e.g., coal, oil and natural gas) for electricity or heat generation, and combustion causes various emissions that may be classified as:

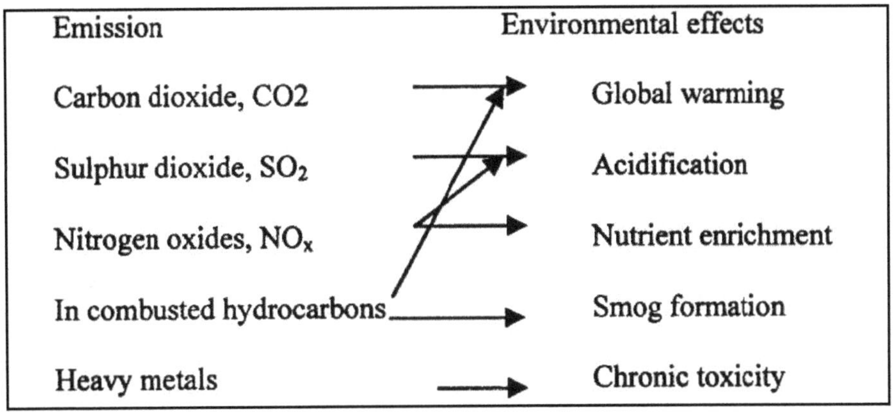

Classification of Emissions.

Characterisation

The life cycle impact assessment includes, as a third element, the characterisation of the inventory data (ISO, 1997 c).

Characterisation is mainly a quantitative step, based on scientific analyses of the relevant environmental processes. It has to assign the relative contribution of each input and output to the selected impact categories. The potential contribution of each input and output to the environmental impacts has to be estimated. For some of the environmental impact categories, there is consensus about equivalency factors to be used in the estimation of the total impact (e.g., global warming potentials, ozone depletion potentials etc.) while for some others there may not be any consensus (e.g., biotic resources, land uses, etc.).

Valuation/Weighting

Characterisation results in a quantitative statement on different impact categories (e.g., global warming, stratospheric ozone depletion and ecotoxicological effects.) Comparison of these categories is not immediately possible. Therefore, the life cycle assessment includes, as a fourth element, a valuation/weighting of the impact categories against each other.

Weighting aims to rank, weight, or if possible, aggregate the results of different life cycle impact assessment categories in order to arrive at the relative importance of these different results. The weighting process is not technical, scientific or objective, as these various life cycle impact assessment results are not directly comparable (e.g., indicators for greenhouse gases or resource depletion.) However, applying scientifically based analytical techniques may assist in weighting. The purpose of weighting the following aspects:

- To express the relative preference of an organisation or group of stakeholders based on policies, goals or aims and personal or group opinions or beliefs common to the group.

- To ensure that process is visible, documentable and reportable.
- To establish the relative importance of the results is based on the state of knowledge about these issues.

Different institutions based on different principles have developed weighting approaches including the following:

- Proxy approach: In this approach, one or a combination of several quantitative measures is stated to be indicative of the total environmental impact. Energy consumption, material displacement and space consumption are examples of using this approach.
- Technology abatement approach: The possibility of reducing environmental burdens by using different technological abatement methods can be used to set a value on the specific environmental burden. This approach can be applied to inventory data as well as impact scores.
- Monetarisation: This approach is based on the premise that utilitarianism (values are measured by the aggregation of human preferences), willingness to pay/accept is an adequate measure of preferences and the values of environmental quality can be substituted by other commodities. This approach can be applied to inventory data as well as impact scores.
- Authorised goals or standards: Environmental standards and quality targets as well as political reduction targets can be used to calculate critical volumes for emissions to air, water, soil or work environment. National or local authorities within a company can formulate the targets or standards.
- Authoritative panels: An authoritative panel can be made up of lay people, experts, representatives of governments, international bodies, etc. The credibility of a panel can be improved by using LCA experts from different societal groups as panellists, peer review sets of valuation criteria and the rules for their application, and a transparent ranking technique, and the documentation of the arguments leading to the final valuation (Volkvein, et. al., 1996).

In the impact assessment step of an LCA, the inventory of chemical and physical releases is translated into an environmental profile and the resources consumed are divided into renewable and non-renewable.

Quantitative Inventories of Environmental Releases and Resources Consumptions for the Production of 1 MJ Heat from Various Fossil Fuels:

Emission type	Unit	Coal	Heavy fuel	Natural gas	Wood	Straw
Carbon dioxide, CO2	g/MJ	96	80	56	98a	99a
Carbon monoxide, CO	mg/MJ	50	10	150	6000	4500
Sulphur dioxide, SO2	mg/MJ	650	500	1	33	110
Nitrogen oxides, NOx	mg/MJ	500	240	350	50	130
Volatile organic compounds, VOC	mg/MJ	15	7	15	1200	-

Polycyclic aromatic Hydrocarbons, PAH	mg/MJ	2.0	0.25	0.13	5000	25
Dioxins, PCDD, PCDF	mg/MJ	0.04	-	0	-	0.3
Arsenic, As	mg/MJ	8	2	0.0	5	-
Cadmium, Cd	mg/MJ	0.9	0.7	0.0	10	-
Chromium, Cr	mg/MJ	4	1.2	0.0	50	-
Mercury, Hg	mg/MJ	1.7	0.1	0.0	1	0.0
Nickel, Ni	mg/MJ	6	400	0.0	-	-
Lead, Pb	mg/MJ	7.6	25	0.0	200	1.0
Particles	mg/MJ	25	1.3	0	45	20
Resource consumption	g/MJ	39	25	23	100	67

The environmental profile shows the potential contribution to all types of environmental effects that are considered important and undesirable. For each individual release listed in the inventory, its potential contributions are calculated and added to the overall contribution to the relevant effect type. Releases that do not contribute to any known effect type will not influence the environmental profile. The list is obviously subject to change due to developments in scientific knowledge, and, to a certain extent, political initiatives. The list should, therefore, be continually adjusted to accommodate such changes. The table gives a list of the environmental effects that ought to be considered in an assessment:

LCA: Environmental Effect Types

Global	Global warming Stratospheric ozone depletion
Regional	Acidification Nutrient enrichment Photochemical ozone formation (smog) Chronic toxicity
Local	Acute toxicity Area degradation, including: risks noise smell Physical disturbances, including: soil erosion deterioration in landscape quality disturbance or destruction of habitats, and hence, ecosystems.

Evaluating Environmental Profiles

Evaluation involves a comparison of the alternatives studied (e.g., products, materials and projects), represented by the environmental, resource and the working environment profiles. The objective of the evaluation is to decide which alternative causes the least environmental damage from a life cycle point of view.

If one alternative is better in all the aspects considered, the evaluation is straightforward. In fact, this is often the case when a product is reviewed first with the purpose of reducing its environmental impacts. However, the evaluation situation gets complex when one alternative is better than the rest in some aspects, while another alternative is better in other aspects. This creates the need to give priorities to the different aspects of the loading profiles. For example, we may ask the questions:

- Which is more important – improving performance in the external environment or in the working environment?
- Is reducing the life cycle contribution to ozone depletion more important than diminishing the release of chemicals causing chronic toxicity?

Clearly, no objective answer can be given to questions like these that involve judgement of fundamentally incomparable items. The setting of the priorities involved in the evaluation step inevitably introduces subjective decisions into the LCA. It should be kept in mind that all the work conducted in LCA-steps leading up to the evaluation must be based on objective and scientific principles. The subjective decisions are hence confined to this last step.

When the inventory is complete and all the emissions listed are classified into environmental impact classes, and converted into common effect units, where possible, the result is an environmental profile for each of the alternatives are examined.

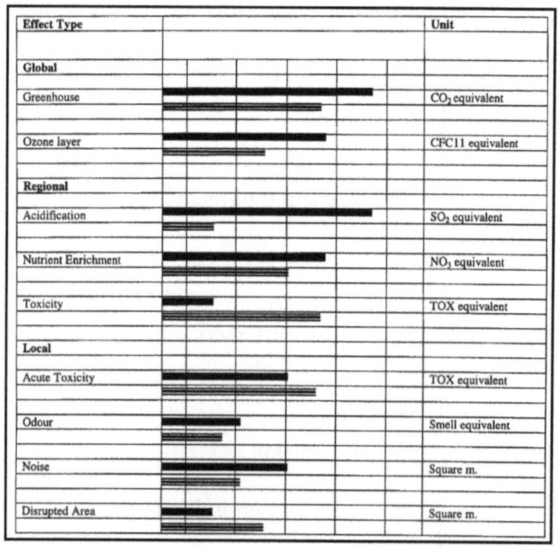

Comparing the profiles, you may find that Alternative 1 contributes less than Alternative 2 to global warming, acidification and nutrient enrichment, while Alternative 2 contributes less to ozone depletion and acute toxicity.

It may be very difficult to weigh between different classes of environmental impacts. But, it will help, if you know which of the differences in contribution to the environmental effects are important and which are not. An example of the kind of question you may have to answer is: How does 15 kg of CO_2 equivalents compare to 3 kg of CFC – 11 equivalents, or to 5 kg of SO_2 equivalents?

To help in the comparisons of this nature, each of the contribution in the environmental profile may be divided by the emission from a suitable common emission scenario. This scenario may, for instance, be the environmental profile of the total emissions of your state or country. In this case, the resulting normalised environmental profile shown indicates the size of the fraction of national emissions due to each of the environmental effect contributions:

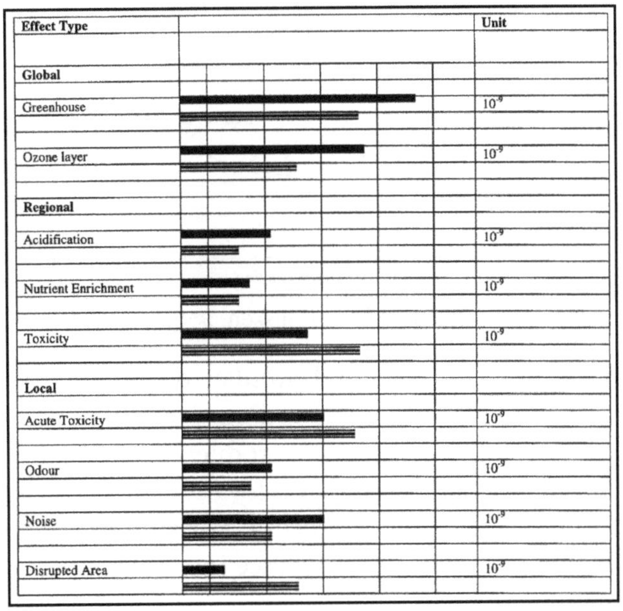

The normalisation step is of particular relevance for comparing alternatives with complicated environmental profiles, i.e., the many different emissions that contribute to different impacts on the environment.

The last step of the LCA is the comparison of the normalised environmental profiles of the different alternatives. The outcome of the evaluation should be the selection of the alternative that performs best from an environmental protection viewpoint.

Different Applications of LCA

Applications for LCA can be broadly classified into private sector applications and governmental applications.

Private Sector Applications

The use of LCA in the private sector varies greatly. To a large extent, this differentiation depends on where a given company is situated in the product chain and the key driver for the LCA activity, e.g., legislation or market competition. For business teams, the LCA tool should be used to understand the environmental issues associated with upstream and downstream processes as well as on-site processes.

Product Development

Using LCA in product development is an obvious choice, as a large part of the future environmental impacts of a product (system) is determined by the design and construction phase. By incorporating LCA in the design phase, companies have the possibility of avoiding or minimising foreseeable impacts compromising the overall quality of the product.

Product development may follow different concepts and routes. The following figure shows some of the common phases of most development methodologies:

Relationships between the Designer's Degree of Freedom and the Level of Information.

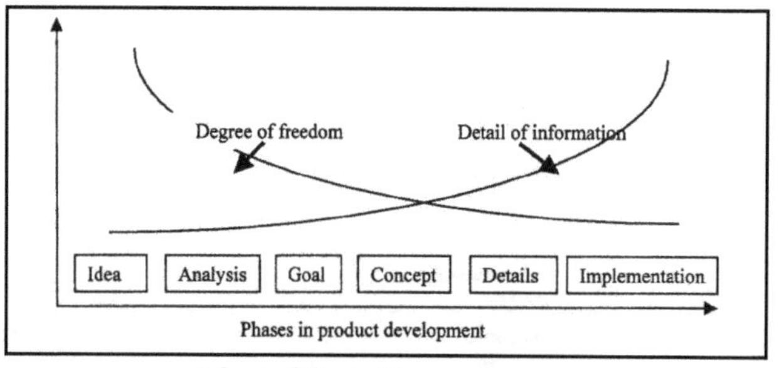

Adapted from Hanssen, 1995.

In the idea-phase, there is almost an unlimited number of possibilities with respect to design, choice of materials, function, etc. The number of options decreases with the development process and changes to the final product and the necessary productive tools often require a whole new development process. It is, therefore, necessary that relevant environmental tools are available and used as early as possible in the development process. For simple products, e.g., packaging, it is possible to apply a detailed and quantitative LCA, since the information on most of the commonly used materials is now available.

Marketing

Marketing is the traditional way of communicating product properties and capabilities, which are consistent with the consumer"s expectations and demands. As the level of environmental consciousness is increasing, the consumer pays more attention to the environmental properties of goods and services. There are four different kinds of environmental marketing. We will describe these, next.

Environmental Labelling (ISO Type-I Labelling)

An environmental label (eco-label) can be seen as a seal of approval for environmentally benign products and can, therefore, be attractive for marketing purposes. Eco-label, at the same time, conveys information to the consumer in a simple but not in an objective way, enabling individuals to include environmental concerns in their own decisions along with considerations for economy and quality.

The general objective of national and supranational eco-labelling scheme is to make products with less environmental impacts visible to the consumer. The success of an eco-labelling scheme, to a certain extent, is dependent on the number of product groups with an eco-label.

Environmental Claims (ISO Type-II Labelling)

An environmental claim is presently defined by ISO (ISO/DIS 14021 and ISO/CD 14022) as a "label or declaration that indicates the environmental aspects of a product or service that may take the form of statements, symbols or graphics on product or packaging labels, product literature, technical bulletins, advertising, publicity or similar applications".

Environmental claims are often uni-dimensional and related to environmental issue of the year or month (e.g., lead-free gasoline, phosphate-free detergent, CFC free hair spray, etc.) Very few of such environmental claims are based on an LCA and, in many cases, focus on irrelevant issues, while failing to address those issues, which are important in a life cycle perspective.

Environmental Declarations (ISO Type-III Labelling)

Environmental declarations may be a tool in eco-marketing to transfer the results from a life cycle investigation of a product (either as a life cycle inventory or a life cycle assessment) to the individual decision-making process of a consumer. The general idea is to give a graphic presentation of a pre-set number of environmental impacts, e.g., by using a bar diagram. Two or more environmental declarations can, on the other hand, illustrate the actual difference between the products and, for environmentally conscious consumers, this can be a valuable supplement to choose between labelled products.

The use of LCA is, thus, a prerequisite for environmental declarations. Standardisation efforts have been initiated by ISO and includes requirements on methodology, transparency, external review, comparative assertions, labelling components, administrative guidelines and procedures governing the accreditation and conduct of type III labelling practitioners.

Organisational Marketing

The classical marketing of environmental performance has mainly been oriented towards products. However, with the increasing number of companies being certified according to ISO 14001, EMAS or BS 7750, some marketing initiatives are being directed towards the environmental capabilities of the companies per se.

As organisations implement the necessary policies for certification, they also encourage formalising the implementation of LCA procedures and life cycle.

Strategic Business Planning

Integration of environmental aspects in strategic business planning is becoming a common feature in many companies. The handling of environmental management schemes such as EMAS (Environmental Management and Auditing Scheme) and ISO 14001 standards. But, many companies still handle the issues on a case-to-case basis.

There are several motivating factors behind the decision to integrate environmental issues, many of which are interrelated. These include, for example, consumer demands, compliance with legislation, community needs for environmental improvement, security of supply and product and market opportunities.

The environmental performance is, thus, changing from being a mandatory property of many products (i.e., all regulatory requirements shall be fulfilled) to being a strong positioning property on the market. An LCA can be used both in relation to existing products and to identify market segments to be opened for environmentally benign products. The LCA information can provide decision-makers with an understanding of the environmental pros and cons of their products and services.

Governmental Applications

Sustainable development has been included as a major item on most of the governmental agendas since the 1992 Rio summit. Although a precise definition of sustainable development has not been given, it is obvious that LCA must be used to ensure that actions towards a more sustainable future will have the desired effect. An LCA, as a specific tool, can ensure this in some cases, while LCA, as an approach or as a strategic tool, can give directions but not the whole answer and must, therefore, be applied along with other tools such as risk assessment, environmental impact assessment, cost benefit analysis and others. The main governmental applications include product oriented policy, deposit-refund schemes, including waste management policies, subsidies and taxation, and general (process-oriented) policies.

Global Warming

This is caused by gases that absorb infra-red radiation (heat) irradiated from the Earth. Increasing their concentrations in the atmosphere may cause the global atmospheric temperature to increase, and this may, through increased melting of polar ice caps and heat-expansion of water in the oceans, lead to rising sea levels and serious climatic changes. The range and the ecological and socio-economic consequences of these effects are impossible to predict at present.

Carbon dioxide, $CO2$, is the most prominent of the global warming gases, contributing about 60% of the total anthropogenic (human- made) atmospheric content of global warming gases. Other important gases are methane, $CH4$, contributing 15% (the main anthropogenic sources are stocks of cattle and cultivation of rice), dinitrogenoxide (or nitrous oxide), $N2O$, with 4% (combustion processes, fertilised agricultural soils) and the halocarbons group, with around 10% (CFCs, HCFCs and halons, all anthropogenic in origin and manufactured for use as solvents, aerosol propellants, heat transmitters, insulators and fire extinguishers).

For carbon dioxide, it is important in the inventory to distinguish between activities that cause a net increase in the atmospheric content (such as combustion of the fossil fuels, oil, coal and gas, or their derivatives such as plastics) and other activities where the CO, $CO2$ contribution would reach the atmosphere anyway, through the biogeochemical carbon cycle. The latter includes the burning of biological materials that would otherwise have decayed naturally and released their carbon content as $CO2$ to the atmosphere. However, the clearance and burning of forests to create agricultural land is an exception from this rule, since the activity causes a net reduction in the global biomass and thus a net increase in the atmospheric content of $CO2$.

Stratospheric Ozone Depletion

The stratosphere is one of the higher strata of the atmosphere (10 - 50 km above the surface of the Earth). It is characterised by the presence of large concentrations of the gas ozone $OCO2$ that constantly exchanges with oxygen through the action of sunlight and various compounds in the stratosphere. The body of ozone in the stratosphere is commonly referred to as the ozone layer. It is of crucial importance for the presence of life as we know it on Earth, since it absorbs most of the ultra-violet (UV) radiations emitted from the sun, which can have very damaging effects, such as causing skin cancers.

It has been found that recent emissions of the man-made halo carbons (described in the paragraph on global warming) are contributing to the breakdown of ozone molecules in the stratosphere, thus favoring ozone degradation over ozone formation and reducing the extent and UV- absorbing capacity of the ozone layer. This may have serious direct effects on life on Earth, especially on the plants and algae that are the primary producers and basis of all food chains. The possible indirect effects are at present hard to assess.

Acidification

Emissions of compounds that contribute to a lowering of the pH in some compartment of the environment are aggregated in this effect type. Acidification mainly occurs in areas where the underground strata are poor in limestone that may neutralise the incoming acids. The biological systems that are affected by acidification are lakes, forests and agriculture. The lowering of the pH in lake water or soil water dramatically changes the living conditions, resulting in massive death of the organisms normally living there (fish, trees, crop plants). Acidification also causes severe economic losses in urban areas, where corrosion of machinery and buildings may be substantially increased.

The contributors to acidification are compounds that are acids themselves (capable of donating protons in chemical reactions) or ones that cause liberations of protons as they experience conversions in a receiving body of the environment. Among the important sources of acidification is combustion of chlorine – containing, fossil fuels, production of metal from sulphide ore, combustion of chlorine substances, e.g., in waste incineration.

Nutrient Enrichment

Nutrient enrichment is the process of enrichment of biotopes (water bodies and soils) in the two main nutrients, nitrogen, N. and phosphorus, P, one of which is generally the nutrient factor limiting biological growth of the system. The effect type incorporates the ensuing changes in chemical and physical living conditions for the organisms and consecutive or cumulative disturbances of the ecosystems. The main focus is on the eutrophication of lakes and coastal waters. Here, the input of nutrients causes algal blooms that shadow out the plant life at the bottom of the water body, even in shallow waters, leading to a vast precipitation of dead algae. The degradation of this additional input of biological material uses up the oxygen dissolved in the water, causing anoxic conditions in the sediments and in the bottom strata. The effect on the fish life and the aquatic ecosystems may lead to dramatic reduction in catches of fish.

Nutrient enrichment of soil ecosystems may cause changes in the species composition of plant communities growing on soils poor in nutrients, thus threatening to extinguish certain plant species.

Through percolation, the nutrient-enriched precipitation may reach the groundwater and cause unwanted concentrations of compounds, particularly of nitrogen, in this source of drinking water.

Contributing to nutrient enrichment are all the compounds containing nitrogen or phosphorus that may be released during their degradation in the environment. Important sources of N and P are agricultural fertilisers and emissions of untreated sewage from urban districts. Combustion processes may also be a significant source of nitrogen.

Photochemical Ozone Formation

Although neurotoxin in themselves, most hydrocarbons will rarely contribute to toxicity in the environment, because their atmospheric half-lives are too short for large concentrations to build up in the troposphere (0 - 10 km above the surface of the Earth). The atmospheric degradation of the hydrocarbons takes place through reactions with nitrogen oxides and hydroxyl radicals present in the troposphere. The reaction schemes include photochemical reactions requiring sunlight, and they lead to the formation of aggressive gaseous compounds, the most important of which is ozone. While the presence of ozone in the higher strata of the atmosphere is indispensable for the existence of life, as we know it, it is highly noxious in the air that we breathe.

Episodes of high concentrations of these reactive gases may occur locally due to emissions of excessive amounts of hydrocarbons, e.g., from traffic during rush hours, but their extent may increase to a regional scale in periods where meteorological conditions prevent the vertical mixing of air within the troposphere. Such episodes are referred to as photo (or photo-chemical) smog.

Ozone is a very reactive molecule, and it will oxidise any reduced compound it encounters including most of bimolecular that are present in the troposphere. Plants in particular suffer from exposure to high ozone concentrations, as they are unable to move during episodes of photo smog, and they depend completely on the exchange of gases with the surrounding air to furnish carbon dioxide for photosynthesis.

Toxicity (Chronic or Acute)

This type of environmental effect includes contributions to toxicity (or poisoning) affecting any kind of living organism or ecosystem (human toxicity as well as environmental toxicity). The toxicity may be acute or chronic, and literally thousands of different chemical compounds may contribute, with toxic effects of many diverse types. Potential contributions to acute or chronic toxicity in the environment are much more difficult to quantify and to aggregate than contributions to any of the other effect types.

Since many different compounds contribute to this environmental effect type, most human activities that might be considered an object for an environmental assessment entail releases that will be toxic in the environment. Therefore, it is difficult to point out individual human activities as the main sources of environmental toxicity. Nevertheless, chemical industries of all types generally cause significant emissions of many different toxins affecting humans and the environment.

Though a quantitative approach to the aggregation or accumulation of toxicity is difficult, this item should under no circumstances be left out of an environmental assessment. Since toxic effects will generally occur on a local scale, the releases contributing to toxicity will often be the ones that are felt most strongly by people living in the surroundings of the project considered. They will, therefore, attract the most attention. Neglecting toxic effects in the classification may strongly reduce the general credibility of the environmental assessment.

Resource Consumption

Apart from reducing the environmental load of the product, a general goal of environmental management of products is to reduce the use of limited or non-renewable resources, so as not to impair the possibilities for future generations on planet Earth. (Current practices should, as a general rule be sustainable.) Therefore, a distinction should be made between renewable and non-renewable resources in the inventory. A renewable resource is defined as one that is being replenished naturally, whereas non- renewable resources are not regenerated at a rate that is significant in a historical perspective.

For historical reasons, use of resources in energy production is considered as a special kind of resource consumption with its own inventory. Also, an inventory of the consumption of energy resources must distinguish between renewable and non- renewable resources.

The bare classification into renewable and non-renewable resources is insufficient for assessment purposes. As a minimum, some further raking of the different kinds of non-renewable resources is necessary. For example, although both sand and copper ore are non-renewable resources, their use should not be considered as equally critical in an LCA. Unfortunately, there is no generally accepted scheme for classification of resources consumption, similar to the scheme adopted for environmental releases.

Life Cycle Analysis Case Study: Substitutes for PVC

After a long, fierce political debate on the desirability and possibility of reducing the consumption of the plastic material PVC, (Poly Vinyl Chloride), The Danish Agency for Environmental Protection commissioned a life cycle assessment of PVC and alternative materials to determine the environmental and health related consequences of substituting other materials for PVC in packaging materials, building materials and office and hospital articles.

The life cycle assessment (LCA) was performed as a screening involving all life cycle stage of the materials, from the extraction of raw materials to their disposal through landfilling, incineration or possible recycling. A more detailed study was performed on what seemed, from the screening, to be the key issues for the life cycle of the material.

This investigation differed from most other LCA studies in assigning equal importance to working environment/health/risk aspects and the external environment.

The system studied was specified by the prevalent conditions in Denmark concerning working environment and safety regulations. The disposal method was considered to be incineration with energy recovery for heating.

The following aspects were studied for each of the materials:

- consumption of resources (including energy),
- potential exposure in the working environment,
- potential effects in the working environment,
- potential exposure in the natural environment,
- potential effects in the natural environment, and risk of accidents.

The materials evaluated were:

- polyvinyl chloride, PVC,
- polyethylene/polypropylene, PE/PP
- polystyrene, PS,
- polyurethane, PUR,
- polyethylene terephthalate, PET,
- vacuum-treated and surface-treated wood,
- paper/cardboard, and
- aluminium.

Several of the investigated materials were only considered as alternatives to PVS in some application. Some conclusions of the study are given below.

Polyvinyl Chloride

The worst problems related to environmental aspects seemed to be:

- emissions of dioxins and chlorinated organic compounds in the waste water from production and combustion gases from incineration of PVC;
- emission of hydrochloric acid and toxic heavy metals (particularly lead and cadmium) that may be added as stabilisers to PVC for some applications.

Those related to health aspects were:

- exposure to carcinogenic vinyl chloride monomer, which may occur during production of the polymer and its manufacture into products;
- exposure to hydrochloric acid and to the carcinogen plasticiser, DEHP, occurring during manufacture of PVC products;
- risk of exposure to chlorine and carcinogenic vinyl chloride monomer in the event of accidents at PVC polymer production facilities.

Polyethylene/Polypropylene

Unlike PVC, these plastics do not contain chlorine, and this leads to substantial improvements in environmental and working environment performance. The energy consumption for production of PE and PP is similar to that required for PVC. The main problem with these materials is that they require addition of flame retardants from some applications, which may introduce environmental problems in the production, manufacture and disposal of the polymers.

The overall conclusion, however, is that the substitution of PVC with polyethylene or polypropylene, where possible, will give substantial improvements in external and working environmental performance.

Polystyrene

The production of polystyrene involves a risk of exposure to the reproduction-damaging substance, styrene. The polymerisation of styrene to polystyrene may involve a risk of explosion. The energy consumption for production of polystyrene is considerably higher than for PVC.

Polystyrene may be expanded into a rigid foam by blowing with gases. In some cases, the gases used are the highly ozone- degrading CFC or HCFC gases. If use of CFCs and HCFCs is avoided, the substitution of polystyrene for PVC will lead to environmental improvement, particularly at the disposal stage. The working environment aspects are unclear.

Polyethylene Terephthalate

PET entails somewhat higher energy consumption than PVC. The overall environmental performance is clearly better for PET than for PVC, both for working environment, external environment and probability and consequences of accidents.

Polyurethane

The production of PUR involves the use of isocyanides, exposure to which is extremely unpleasant in the working environment during production of the polymer and manufacture into products.

PUR may, similarly to polystyrene, be expanded by blowing with CFCs or HCFCs. If this application is disregarded, the potential effects of PUR in the external environment seem smaller than for PVC, particularly at the disposal stage (due to the absence of chlorine and toxic metals). For working environment conditions, however, there is some aggravation compared to PVC. Overall, the report does not recommend a substitution of PVC with PUR.

Vacuum-treated and Surface-treated Wood

In contrast to all the polymer materials, wood is a renewable resource and generally, little or no fossil energy is required for its production. The manufacture of vacuum-treated wood, e.g., into window frames, involves the exposure of works to carcinogenic wood dust, and the vacuum-treatment of the wood entails emissions of heavy metals into the external environment.

Nevertheless, the overall conclusion qualifies vacuum- or surface- treated wood as an environmentally advantageous alternative to PVC.

Paper/Cardboard

Like wood, paper and cardboard are made from renewable resources. Their energy consumption is higher than for wood, but still lower than for PVC.

The environmental profile of paper and cardboard depends strongly on the process used. Often, the pulp is bleached using chlorine compounds, resulting in emission of chlorinated organic compounds (including dioxins) in the wastewater. Generally, pulp production involves emissions of large quantities of toxic and oxygen-consuming organic material in the wastewater that is worse than the loading from the PVC-production. The disposal of paper or cardboard after its use is substantially better from an environmental point of view, and overall, paper or cardboard is recommended over PVC for packaging purposes and similar applications.

Aluminium

Production of primary aluminium from bauxite is extremely energy- consuming and involves some serious working environment problems and hazards. Together these aspects disqualify the substitution of PVC with primary aluminium.

For secondary aluminium, i.e., aluminium recycled through collection and recasting, the situation is different since it may be recycled, and the loading of the original production of primary aluminium reduced correspondingly. The preference of PVC or aluminium thus depends on the efficiency of the recycling of aluminium.

Life Cycle Analysis: A Case of Steel Sector in India

Since 1999, with an objective of optimum utilisation of resources, Life Cycle Assessment Study in the Steel Sector has been continued with joint funding of the Ministry and three identified steel plants, viz, Bhilai Steel Plant, SAIL; Jamshedpur Steel Plant of TISCO and Vizag Plant of Rashtriya Ispat Nigam Ltd. (RINL). The salient findings of the study are summarised herein.

Scope of the Study

- Cradle to Gate (RM mining to Crude Steel).
- Includes Inventory Analysis.
- Impact assessment.
- Improvement Analysis.

Boundary of the Study

- Crude Steel at Factory Gate.
- Quantification Unit - Kg of crude steel at factory gate.
- Includes.
- Steel Making.
- Production & Transportation of RM.

- Production & Transportation of Energy.
- Production & Transportation of Consumables.

Data Collection & Validation

- Main Source of Data - Annual Statistics.
- Coke Oven Gas, BF Gas and BOF Gas conversion from calories to joules by 4.1868 joules/cal.
- Heat Content of 8 ata steam 3000 MJ/tonne and 18 ata 3260MJ/tonne.
- Fugitive Dust and drinking water not included.
- Data Validation for FY1997 & FY 1998 by MECON Primary data validation by log sheet checks.
- Secondary data validation - Monthly, Quarterly, Half Yearly &annual Audits.
- Energy Audits - Steam, Electricity, Fuel.
- Emission Audits - Weekly Stack & effluent analysis.
- Performance Audits - Daily Production Target monitoring at department level and against techno-economic parameters at plant level.
- Technology Audits.

Results

Sl. No.	Item	World Av. (Kg/Kg of steel)	Units Av. (Kg/Kg of steel)
1.	Coal	0.536	0.989
2.	Dolomite	0.0291	0.158
3.	Limestone	0.0001	0.171
4.	Iron Ore	1.38	1.813
5.	Steel Scrap	0.142	0.110
6.	Water	8.1(l/kg)	4.74 (l/kg)
7.	Energy	21.7 (MJ/kg)	29.77 (MJ/Kg)

Emission Results

Sl. No.	Item	World Av.(g/Kg of steel)	Units Av (g/Kg of steel)
1.	CO_2	1734	2924.65
2.	CO	26.4	63.09
3.	NOx	1.57	1.39
4.	SOx	1.48	1.70
5	Particulates	1.57	1.275

Result Analysis

- Ash content of Domestic coal high so coal consumption high.
- Alumina in Domestic ore high so higher Dolomite requirement.
- Iron Ore lump used in cooling in BOF in India nowhere else in the world so higher Iron Ore Consumption.
- Steel plants outside India mostly on shores, don't recycle water, hence high consumption.

- World over steel plants use BF gas top pressure to extract electricity not in this unit.
- BF gas having higher Carbon though lower calorific value compared to other gases used so higher CO_2 emission.
- World over alternate fuels used in BF like oil, Natural Gas, Pulverised Coal etc. BF gas generated per tonne is low.
- High CO emission in Sinter making process.
- Low Sulphur in Indian Coals so low SOx emission.
- ODS use is low.

Impact Assessment

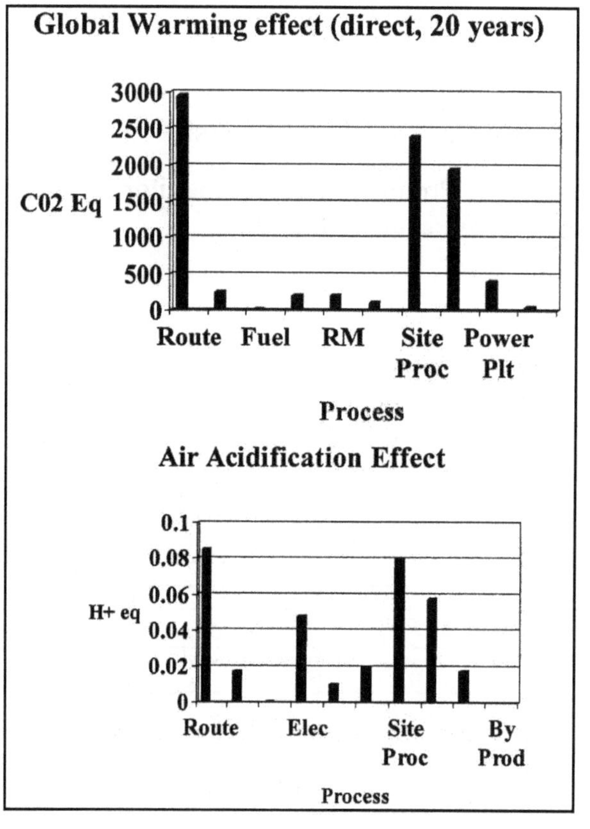

Improvement Planning

- Increase in proportion of imported coal in coal blend.
- New Sinter machine installation.
- Installation of Top Gas Pressure Recovery to harness electricity.
- Coal Dust Injection in Blast Furnace.
- Increased Recovery of BOF gas.

References

- "ISO 45001 - Occupational health and safety". ISO.org. International Organisation for Standardization. Retrieved 20, May 2020
- Sciubba, E (2004). "From Engineering Economics to Extended Exergy Accounting: A Possible Path from Monetary to Resource-Based Costing" (PDF). Journal of Industrial Ecology. 8 (4): 19–40. doi:10.1162/1088198043630397
- Finnveden, G., Hauschild, M.Z., Ekvall, T., Guinée, J., Heijungs, R., Hellweq, S., Koehler, A., Pennington, D. & Suh, S. (2009). Recent developments in Life Cycle Assessment. Journal of Environmental Management 91(1), 1-21
- Curkovic, S, and Sroufe, R.P., "Using ISO 14001 to Promote a Sustainable Supply Chain Strategy," accepted, International Journal of Business Strategy and the Environment, Vol. 20, No. 2, 71-93, 2011
- Jiménez-González, C.; Kim, S.; Overcash, M. Methodology for developing gate-to-gate Life cycle inventory information. The International Journal of Life Cycle Assessment 2000, 5, 153–159
- "PAS 2050:2011 Specification for the assessment of the life cycle greenhouse gas emissions of goods and services". BSI. Retrieved on: 25, April 2020
- S. Singh; B. R. Bakshi (2009). "Eco-LCA: A Tool for Quantifying the Role of Ecological Resources in LCA". International Symposium on Sustainable Systems and Technology: 1–6. doi:10.1109/ISSST.2009.5156770. ISBN 978-1-4244-4324-6
- Curran, Mary Ann. "Life Cycle Analysis: Principles and Practice" (PDF). Scientific Applications International Corporation. Archived from the original (PDF) on 18 October 2011. Retrieved 24, August 2020
- Hendrickson, C. T., Lave, L. B., and Matthews, H. S. (2005). Environmental Life Cycle Assessment of Goods and Services: An Input–Output Approach, Resources for the Future Press ISBN 1-933115-24-6
- Brinkman, Norman; Eberle, Ulrich; Formanski, Volker; Grebe, Uwe-Dieter; Matthe, Roland (15 April 2012). "Vehicle Electrification - Quo Vadis". VDI. Retrieved 27, April 2020

Permissions

All chapters in this book are published with permission under the Creative Commons Attribution Share Alike License or equivalent. Every chapter published in this book has been scrutinized by our experts. Their significance has been extensively debated. The topics covered herein carry significant information for a comprehensive understanding. They may even be implemented as practical applications or may be referred to as a beginning point for further studies.

We would like to thank the editorial team for lending their expertise to make the book truly unique. They have played a crucial role in the development of this book. Without their invaluable contributions this book wouldn't have been possible. They have made vital efforts to compile up to date information on the varied aspects of this subject to make this book a valuable addition to the collection of many professionals and students.

This book was conceptualized with the vision of imparting up-to-date and integrated information in this field. To ensure the same, a matchless editorial board was set up. Every individual on the board went through rigorous rounds of assessment to prove their worth. After which they invested a large part of their time researching and compiling the most relevant data for our readers.

The editorial board has been involved in producing this book since its inception. They have spent rigorous hours researching and exploring the diverse topics which have resulted in the successful publishing of this book. They have passed on their knowledge of decades through this book. To expedite this challenging task, the publisher supported the team at every step. A small team of assistant editors was also appointed to further simplify the editing procedure and attain best results for the readers.

Apart from the editorial board, the designing team has also invested a significant amount of their time in understanding the subject and creating the most relevant covers. They scrutinized every image to scout for the most suitable representation of the subject and create an appropriate cover for the book.

The publishing team has been an ardent support to the editorial, designing and production team. Their endless efforts to recruit the best for this project, has resulted in the accomplishment of this book. They are a veteran in the field of academics and their pool of knowledge is as vast as their experience in printing. Their expertise and guidance has proved useful at every step. Their uncompromising quality standards have made this book an exceptional effort. Their encouragement from time to time has been an inspiration for everyone.

The publisher and the editorial board hope that this book will prove to be a valuable piece of knowledge for students, practitioners and scholars across the globe.

Index

K

M

N

O

P

R

U

W

CPSIA information can be obtained
at www.ICGtesting.com
Printed in the USA
LVHW060855030122
707669LV00002B/182

9 781639 890446